Defect Sizing Using Non-destructive Ultrasonic Testing

Wolf Kleinert

Defect Sizing Using Non-destructive Ultrasonic Testing

Applying Bandwidth-Dependent DAC and DGS Curves

Springer

Wolf Kleinert
Bonn
Germany

ISBN 978-3-319-32834-8 ISBN 978-3-319-32836-2 (eBook)
DOI 10.1007/978-3-319-32836-2

Library of Congress Control Number: 2016937519

Printed on acid-free paper

This Springer imprint is published by Springer Nature
The registered company is Springer International Publishing AG Switzerland

Preface

In 1982 I started my career in the Krautkrämer Company in Cologne. Since then I have worked in the field of non-destructive material testing using ultrasonics up to my retirement in the end of 2014. In the past couple of years I was particularly engaged with the distance–gain–size (DGS) method for the sizing of reflectors.

This activity started when a colleague of mine, Michael Berke, came to me showing the result of a software test. The software tested was the implementation of the DGS method in an ultrasonic Flaw Detector. The software test showed strange deviations using the DGS method applied to measurements with an angle beam probe. The first approach, assuming a software bug, had to be abandoned quickly. This led to the development of new innovative angle beam probes, single element as well as phased array probes. Many iterations were necessary during this development. The prototypes were improved step-by-step until the result was satisfactory. In Chaps. 4–6 the development is described in great detail.

A lot of insight was gained during the years of this development, finally resulting in probes which are fully modeled. The sound fields of these probes can be calculated easily. Due to this fact, bandwidth-dependent DGS curves, respectively DAC curves for flat-bottomed holes and side-drilled holes could be engineered.

GE Sensing & Inspection Technologies GmbH in Huerth, Germany applied for several patents covering these new probes including the bandwidth-dependent DGS and DAC curves. When referring to these probes the term *true*DGS® will be used which is a registered trademark of GE Sensing & Inspection Technologies GmbH.

Bonn
April 2016

Wolf Kleinert

Preface

Acknowledgments

Without the help and assistance of my former colleagues the development of the *true*DGS® probes would have been impossible. Particularly with Gerhard Splitt I spent a lot of time discussing the next steps in the development. Additionally he checked all my equations and formulas for the calculation of these probes. In the patent application of the *true*DGS® probes both of us are mentioned as the inventors. Furthermore I appreciate the work of York Oberdoerfer and his team. They built these probes and were very patient following all the necessary iteration steps. It is hard to imagine how many measurements using test blocks were needed to validate the new probes. This large number of measurements was carried out by York and his team. York and I have had lots of technical discussions which all helped to improve the new probes.

Last but not least, I thank my wife Brigitte who patiently accepted my mental absence during the writing of this book which was sometimes connected with a bad mood. Additionally I thank her and my sons for proofreading this document several times.

Contents

List of Figures

List of Tables

Chapter 1
Introduction

Abstract Defect sizing nondestructively using ultrasonics is a very complex task. In the daily routine of ultrasonic operators, defect sizing is done by comparison of the echo amplitude of a natural defect to the amplitude generated by an artificial defect at the same distance from the transducer as the natural defect. The artificial defects used are not the same globally. In parts of the world, influenced by the USA, side-drilled holes are used while under European influence flat-bottomed holes are preferred. Both procedures are described in standards and codes. When side-drilled holes are used, a distance amplitude correction curve is recorded. For flat-bottomed holes, the so-called distance–gain–size method is applied.

Defect sizing nondestructively using ultrasonics is a very complex task. The resulting echo amplitude of a defect depends on many possible influences such as

- size of the defect
- surface condition of the defect
- orientation of the defect to the sound beam
- etc.

There are different approaches to solve this problem or at least get a better understanding of the defect under test. Just two examples are mentioned here **SAFT**: (Synthetic Aperture Focusing Technique) and **TFM** (Total Focusing Method). Both methods require complex set-ups and high computer power.

Different sizing methods are used for the daily work of an ultrasonic operator. In this area, speed of testing and the reproducibility of the test results are key. Therefore, already in the early stages of ultrasonic testing, the reflectivity of a natural defect was compared to an artificial defect at the same sound path. Although we are living in a more global environment, natural defects are compared to different artificial defects depending on the area on the globe where the sizing is performed. When the sizing is done in an area influenced by the USA, the size of a defect is given by the diameter of a side-drilled hole (SDH) in the same depth generating the the same echo amplitude as the natural defect. If the sizing is performed in a part of the world influenced by Europe, the defect size will be given by the diameter of a flat-bottomed hole (FBH) which generates the same echo amplitude if hit perpendicularly by the sound beam and being in the same distance from the transducer as the natural defect.

W. Kleinert, *Defect Sizing Using Non-destructive Ultrasonic Testing*,
DOI 10.1007/978-3-319-32836-2_1

Whenever side-drilled holes are used for comparison, the so-called Distance Amplitude Correction (DAC) curves are applied while the natural defect is compared to the diameter of a flat-bottomed hole (FBH) the distance–gain–size (DGS) method is utilized. The DAC method is based on measurement values while the DGS method is based on a theoretical approach [1, 2]. In the following, a short description of these two different approaches is given.

1.1 Distance Amplitude Correction Curve

A DAC curve needs to be recorded for each probe and each angle to be used for sizing. A reference block with side-drilled holes is required for recording a DAC. This block is ideally manufactured from the same material as the specimen under test. Figure 1.1 shows a sketch of such a test block.

For recording a DAC curve, the echo from each side-drilled hole has to be maximized and the top of the echo has to be marked on the screen of the ultrasonic instrument used, while maintaining a constant gain setting. Today's ultrasonic instruments have functions to support the recording of these curves. Figure 1.2 shows a resulting DAC curve. If the task is to test, e.g., a weld with different angles with 45°, 60°, and 70°, for all these angles a DAC curve needs to be recorded which is quite time consuming. If phased array angle beam probes are used, the task of recording gets very time consuming since for each angle the recording has to be performed. The reason is that phased array angle beam probes change the near field length as well as the delay line and the sound exit point in the wedge with the angle. In addition, the reference blocks required are quite expensive.

The requirements for this sizing method is described by different organizations, such as **ASME** (American Society of Mechanical Engineers), **ASTM** (American Society for Testing and Materials) and **AWS** (American Welding Society).

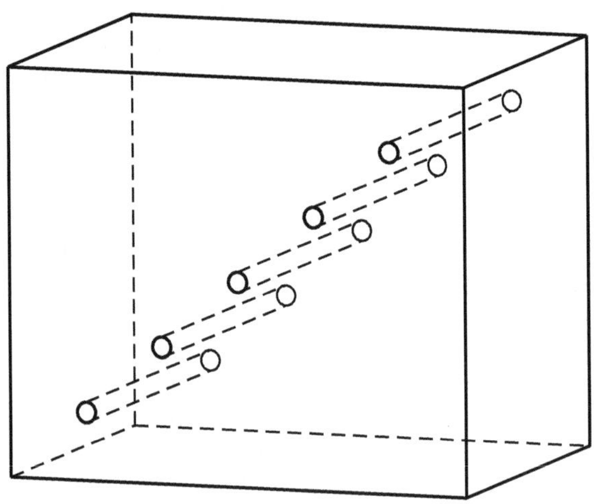

Fig. 1.1 Sketch of a test block with side-drilled holes in different depths

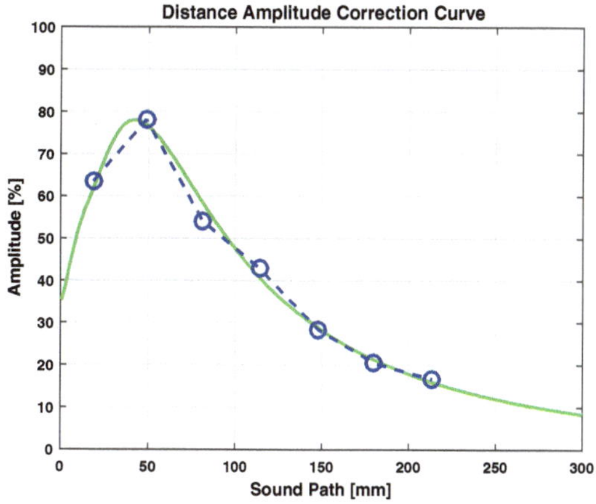

Fig. 1.2 Recorded distance amplitude correction *curve*

1.2 Distance-Gain-Size Method (DGS)

The DGS method has been developed in the late 1950s of the past century by the Kraukrämer brothers [2] for straight beam probes with flat spherical transducers. The so-called general DGS diagram shows the interdependencies of the probe used and the resulting echo amplitude from different sizes of flat bottom holes in different depths. Figure 1.3 shows the general DGS diagram taken from the standard EN ISO 16811:2012[1] [3]. The diagram contains one curve showing the gain differences for a back wall in different depth. The other curves represent the dB differences for flat-bottomed holes with different diameters. The diameter of the flat-bottomed holes are given as fraction of the diameter of the transducer used. The y-axis shows the dB differences and the x-axis shows, in a logarithmic presentation, the distance between transducer and flat-bottomed holes, respectively, to the back wall in multiples of the near field length.

The interdependencies in the far field are characterized by straight lines in the logarithmic presentation. These lines have been calculated theoretically. The interdependencies in the range of the first few near field lengths have been derived using measurements in water as given in [1]:

> ...and experimental measurements made in the sound field between. Measurements were made with circular-disc reflectors in water.

A so-called special DGS diagram for a certain probe can be derived from the general DGS diagram. In today's ultrasonic instruments, the general DGS diagram is stored and the special diagram for the probe used is calculated by the instrument. It is sufficient to record the echo from a flat back wall as reference for sizing. For each amplitude during testing, the instrument displays the so-called Equivalent Reflector Size (ERS). The ERS equals the diameter of a flat-bottomed hole which would

[1]Reproduction with permission of DIN Deutsches Institut für Normung e. V.

ISO 16811:2012(E)

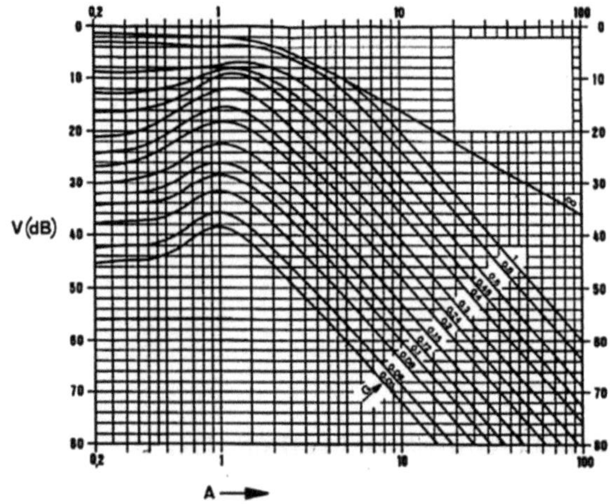

Key

A normalized distance
V gain in dB
G normalized reflector size

Fig. 1.3 General DGS Diagram taken from EN ISO 16811:2012 [3]

generate the same amplitude in the given distance if hit perpendicularly by the sound beam. If a sound attenuation has to be considered, the values for the sound attenuation have to be known and taken into account. In Chap. 2, the DGS method will be discussed in detail.

1.3 Key Differences: DAC Versus DGS

The key differences between the reflectivity of side-drilled and flat-bottomed holes are best described in the far field of the probe used. Doubling the diameter of a flat-bottomed hole results in a gain difference of 12 dB; doubling the diameter of a side-drilled hole results in a gain difference of 3 dB. Using constant diameters and doubling the sound path will result in a 12 dB difference for flat-bottomed holes, while with side-drilled holes the difference equals 9 dB.

1.4 Foresight to This Book

A brief introduction to the state of the art of using DGS will be given. The deviations in sizing using DGS with conventional angle beam probes will be reported. A new angle beam probe technology will be introduced. The sizing using DGS is significantly more precise with these probes. These probes can be handled mathematically much easier than conventional probes.

For these probes, DAC curves can be calculated [4]. Therefore, it is sufficient to record a DAC curve for one single angle alone. The curves for all other angles will be calculated. This is a huge productivity gain particularly when phased array angle beam probes are used which cover a large range of angles. This new probe technology enables the calculation of DGS and DAC curves in the entire sound field range even bandwidth dependently. With this approach, DGS and DAC procedures are getting more similar to each other. Maybe this will lead finally to a global approach of sizing ultrasonically.

References

1. Krautkramer, J., Krautkramer, H.: Ultrasonic testing of materials. In: 4th Fully Revised Edition Translation of the 5th Revised German Edition, Springer (1990)
2. Krautkrämer, J.: Fehlergrössenermittlung mit Ultraschall, Arch. Eisenhüttenwesen 30 (1959)
3. Ultrasonic testing—Sensitivity and range setting (ISO 16811:2012), German version EN ISO 16811:2014
4. Kleinert, W., Oberdörfer, Y.: Calculated bandwidth dependent DGS and DAC curves for phased array sizing. In: Proceedings of ECNDT Prag (2014). http://www.ndt.net/events/ECNDT2014/app/content/Paper/165_Kleinert.pdf

Chapter 2
State of the Art: DAC and DGS

Abstract A brief introduction of the recording of DAC curves is given. The DGS method is presented in more details. The development of the DGS method is described. How to derive a special DGS diagram for a certain probe from the general DGS diagram is explained as well as the necessary adaptations for sound attenuation and other influences. Examples for sizing using the DGS method is given for a straight beam probe and for an angle beam probe.

The state of the art of the two techniques DAC and DGS will be discussed in the following. Using DAC curves does not need a lot of explanation. The understanding of the distance–gain–size (DGS) method will require more details.

2.1 Distance Amplitude Curve

As already described in the introduction of this book, recording of a DAC curve is straightforward. Each echo of each side-drilled hole in the reference block has to be maximized and the echo peak has to be marked on the screen of the ultrasonic instrument keeping the gain setting constant. The marked echo peaks are connected by a line using an appropriate pen. Today's ultrasonic instruments have functions providing help for recording the DAC curve and the curve will be displayed on the screen electronically. As mentioned before, the reference block is ideally manufactured from the same material as the specimen under test. In this case, the material characteristics such as sound attenuation and absorption are taken into account automatically. The disadvantages of this method are the cost of the reference blocks and the time-consuming recording procedure.

W. Kleinert, *Defect Sizing Using Non-destructive Ultrasonic Testing*,
DOI 10.1007/978-3-319-32836-2_2

2.2 Distance–Gain–Size Method

The DGS method has been developed in 1959 by the Krautkrämer [1] brothers for flat circular transducers. In the far field, the general DGS diagram has been calculated theoretically [2] while in the range of the first few near field lengths, measurements were taken to define the curves. The calculated curves are straight lines in the logarithmic presentation of the DGS diagram. Following is an important quotation from the Krautkrämer book [1]:

> ...since the local fluctuations in the near field depend quite sensitively on the pulse length and the transmitter design. In the intermediate range therefore the general DGS diagram can only give approximate results but for a particular transmitter design a special DGS diagram can of course be established by experiment.

Figure 2.2 is a digitized version of the general DGS diagram published in EN ISO 16811:2012 [3], Fig. 1.3. Comparing Figs. 2.1 and 2.2 shows how important this quotation is. The DGS diagram is in the near field heavily dependent on the bandwidth of the probe. Later in this book, the bandwidth-dependent calculation of the DGS diagram for the entire range of sound paths will be discussed in detail.

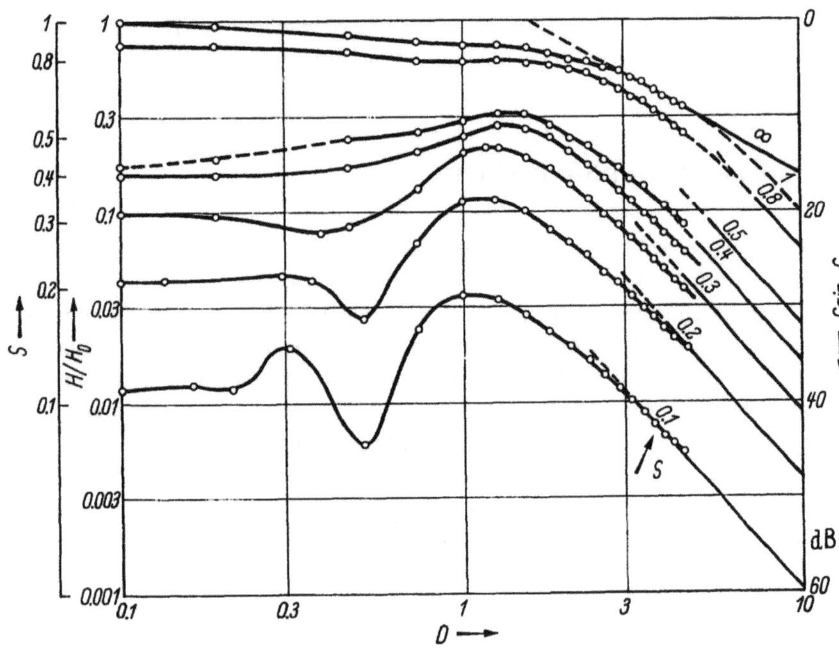

Fig. 2.1 Figure from the Krautkrämer book [1]

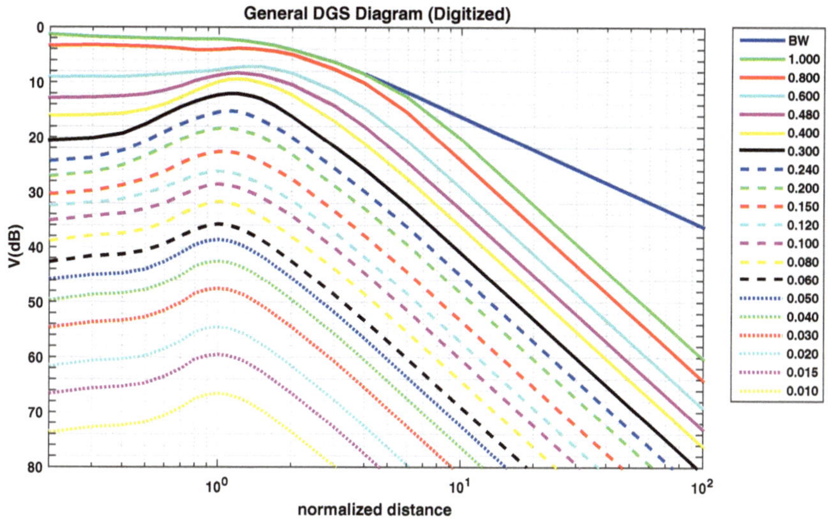

Fig. 2.2 Digitized general DGS diagram

2.2.1 EN ISO 16811:2012

The dependency of the DGS diagram on the bandwidth seems to be forgotten. In the standard EN ISO 16811:2012 [3], a general DGS diagram is published, Fig. 1.3. The bandwidth dependency is not mentioned in detail. But the use of the DGS method is limited to sound paths larger than 0.7 near field lengths, presumably to avoid deviations based on the bandwidth-dependent variations in the near field.

It was not known to the author when the general DGS diagram published in the EN ISO 16811:2011 was developed. Presumably, the development of this diagram was at a time when different circumstances were valid

- Mainly narrow band probes were used.
- A resolution of 0.1 dB for gain setting was not available.
- At this time, the equivalent reflector size was derived by manual interpolation in the logarithmic scale. Today, ultrasonic instruments calculate the equivalent reflector size and display it with a resolution of a tenth of a millimeter.

Therefore, deviations in the evaluation, particularly in the near field and in the intermediate range, were not detected or even accepted.

2.2.2 DGS Evaluation

In this section, the method how to derive a special DGS diagram for a certain probe from the general DGS diagram will be discussed. To fulfill this task, the distances on the x-axis in the general DGS diagram have to be multiplied with the near field length of the probe used. In addition, the size indication (G in the general DGS diagram, Fig. 1.3) has to be multiplied with the diameter of the transducer.

2.2.2.1 Straight Beam Probe

As mentioned in the introduction, the DGS method has been developed for straight beam probes with flat spherical transducers [2]. For this example, a probe with the following parameters will be used:

- frequency: $f = 2\,\text{MHz}$
- diameter: $D = 10\,\text{mm}$
- sound velocity in the test material: $c = 5{,}920\,\text{m/s}$
- the length of the delay is negligible.

The first step is to calculate the near field length N of the probe using the effective diameter D_{eff}, with $D_{eff} = 0.97\,D$, utilizing the following formula:

$$N = \frac{D_{eff}^2 - \lambda^2}{4\,\lambda} \approx \frac{D_{eff}^2}{4\,\lambda} \tag{2.1}$$

where λ is the wavelength.

To derive the special diagram for this probe, the digitized general DGS diagram is taken, refer Fig. 2.2. The values of the x-axis are multiplied with the near field length N and the size indication G is multiplied with the diameter D of the transducer. The result is shown in Fig. 2.3.

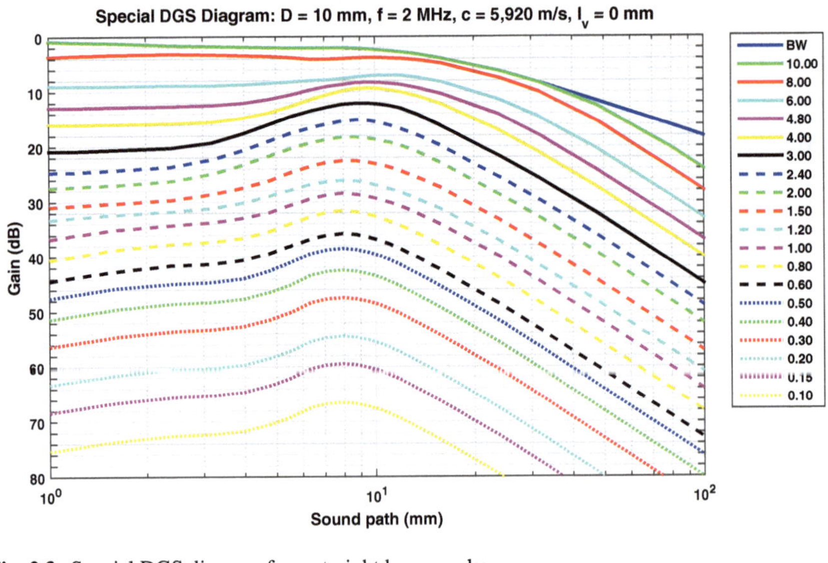

Fig. 2.3 Special DGS diagram for a straight beam probe

To evaluate a reflector, applying the DGS method, a reference echo is required. In this example, the reference echo is taken from the planar back wall of a 40 mm thick test block. Let the gain setting needed to have this reference echo at 80 % screen height be $G_r = 16.8\,\text{dB}$. A reflector is detected at a sound path of 20 mm. This echo as well is set to 80 % screen height with a gain setting of $G_d = 44.2\,\text{dB}$. The gain difference ΔG of these two gain settings is calculated

$$\Delta G = G_d - G_r = 44.2\,\text{dB} - 16.8\,\text{dB} = 27.4\,\text{dB} \tag{2.2}$$

For the evaluation using the DGS method, a point at a sound path of 40 mm on the back wall curve is marked. A second point is marked at ΔG below the first point on the back wall curve. A parallel to the x-axis through the second point is drawn up to the intersection with a line perpendicular to the x-axis at a sound path of 20 mm. This intersection point is the result of the DGS evaluation. At the curve with the intersection point, the equivalent reflector size can be read. In the example given, the equivalent reflector size (ERS) is 1.0 mm, Fig. 2.4. If the intersection point is between two curves, an interpolation between these two curves is required.

2.2.2.2 Angle Beam Probe

As mentioned before, the DGS method was developed for straight beam probes with planar spherical transducers. But, later on, the DGS method was as well applied to angle beam probes [1]:

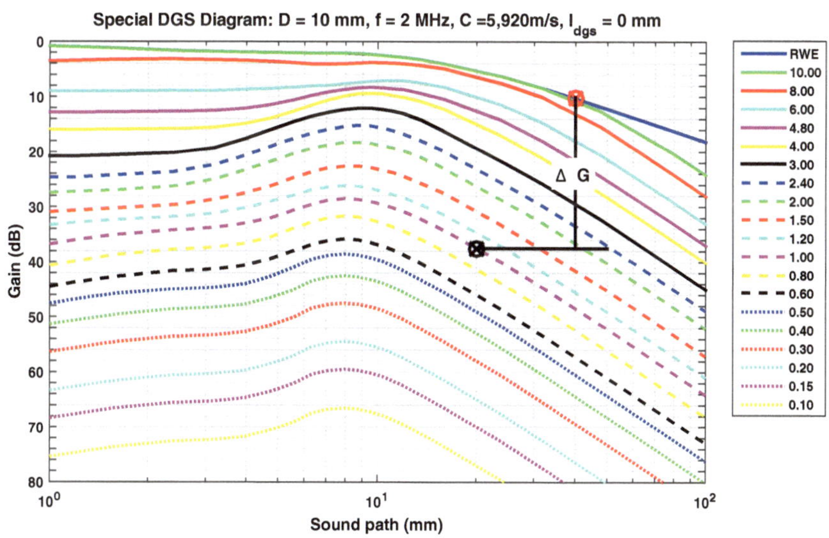

Fig. 2.4 DGS evaluation for a straight beam probe

DGS diagrams have also been established for transverse waves when used with so-called angle probes, cf. Chaps. 19 and 20.

In Chap. 3, it will be seen that the DGS method can lead to oversizing reflectors when conventional angle beam probes are used. First, the state of the art of the DGS evaluation using angle beam probes will be discussed. Usually, angle beam probes have rectangular transducers. With angle beam probes, the DGS evaluation is a bit more complex

- For the rectangular transducer a so-called equivalent circular transducer has to be determined.
- The delay line is not negligible.
- With transverse waves, normally, the sound attenuation has to be taken into account.
- The reference echo is usually taken from the arc of the calibration standard K1 or K2. In this case, the amplitude correction value ΔV_k has to be considered because the arc has a different reflectivity as a flat back wall.
- When using higher frequencies, the sound attenuation in the calibration standard has to be taken into account as well.
- If the surface qualities of the calibration standard and the specimen under test are different, a transfer correction ΔV_t has to be applied.

For calculating the equivalent circular transducer, a correction value based on the side ratio of the rectangular transducer is required [1], refer to Table 2.1. In this table, the following identifiers are used:

- a: half of the larger side of the rectangular transducer
- b: half of the smaller side of the rectangular transducer
- h: correction value

The near field length N of the angle beam probe can be calculated according to the state of the art

$$N = h\,\frac{a^2}{\lambda} \tag{2.3}$$

Table 2.1 Correction values for rectangular transducers [1]

Ratio of sides b/a	Correction value h
1.0	1.37
0.9	1.25
0.8	1.15
0.7	1.09
0.6	1.04
0.5	1.01
0.4	1.00
0.3	0.99
0.2	0.99
0.1	0.99

with λ being the wave length in the test material. Now, using Eq. (2.1), the diameter D of the equivalent circular transducer can be derived

$$D \approx \sqrt{4\,\lambda\,N} \tag{2.4}$$

According to the state of the art, the delay length has to be converted using the so-called near field equivalent

$$l_{dgs} = l_p\,\frac{c_d}{c_m} \tag{2.5}$$

with:

- l_p: physical delay in the wedge of the probe
- l_{dgs}: delay to be considered when deriving the special DGS diagram
- c_d: sound velocity in the wedge of the probe
- c_m: sound velocity in the test material

With the known frequency f of the probe, all values needed to derive the special DGS diagram are now known.

As an example, an angle beam probe with the following data will be used:

- frequency: 4 MHz
- transducer: $8 \times 9\,\text{mm}^2$
- delay: 7 mm
- sound velocity in the wedge 2.73 km/s

The sound velocity in the test material is 3.255 km/s. The probe delay l_p is converted using the ratio of the sound velocities resulting in $l_{dgs} = 5.9\,\text{mm}$. For the near field length, the result is $N = 30.8\,\text{mm}$ and the diameter of the equivalent circular transducer yields $D = 10\,\text{mm}$. The resulting special DGS diagram is represented in Fig. 2.5.

Note:

The DGS diagram is not corrected, for example, the sound attenuation or any other influence. All possible influences have to be corrected manually by adapting the measured dB values accordingly. The following additional parameters are required for the DGS evaluation for angle beam probes:

- ΔV_k: The amplitude correction value specifies by how many dB the echo from the arc of the calibration standard used is higher than the echo from a flat back wall at the same sound path. This value can be taken from the data sheet of the used probe.
- κ_k: Sound attenuation in the calibration standard. This value can be estimated or better measured.
- κ_m: Sound attenuation in the test material. This value has to be measured.
- ΔV_t: transfer correction. This value is required to compensate for different surface qualities of the calibration standard and the test specimen. This value needs to be measured.

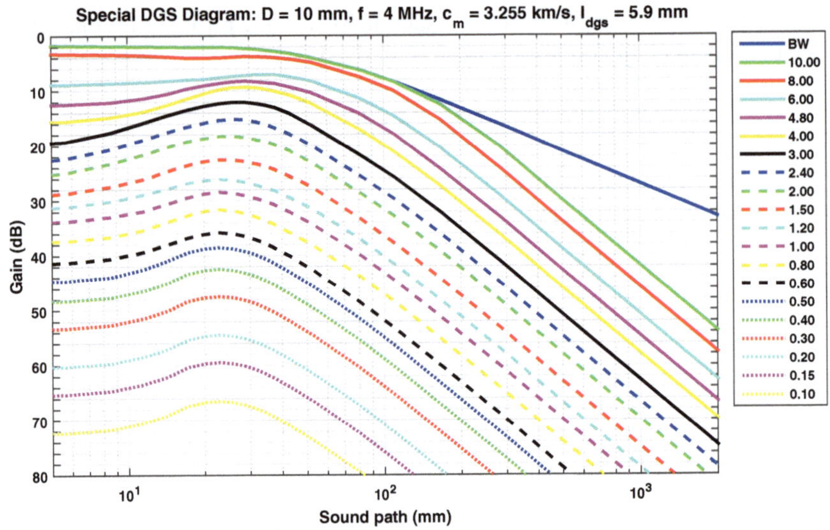

Fig. 2.5 Special DGS diagram for an angle beam probe

An example is given deriving the special diagram for the used probe. With this probe, the reference echo is taken from the 25 mm arc in the calibration standard K2. In the data sheet of this probe, the amplitude correction value ΔV_{K2} is given.

Measurement of the Sound Attenuation in the Calibration Standard:

For measuring the sound attenuation in the calibration standard, the K1 block will be utilized. This can be done because both calibration standards K1 and K2 are made from the same material. To determine the sound attenuation both a V and a W through transmission is measured. In both cases, the echo amplitude is set to 80 % screen height and the necessary gain settings are noted. This measurement must be taken using a pair of probes equivalent to the probe used for the DGS evaluation. Figure 2.6 illustrates the V and W through transmission. The thickness of the calibration standard K1 is $d = 25$ mm. The sound paths s_v for the V through transmission and s_w for the W transmission are calculated from the angle of incidence β and d

$$s_v = \frac{d}{\cos\beta} \tag{2.6}$$
$$s_w = \frac{2\,d}{\cos\beta}$$

For the example with the angle of incidence $\beta = 60°$ the sound paths result in $s_v = 50$ mm and $s_w = 100$ mm. These sound paths are marked on the back wall curve in the special DGS diagram of the probe used and the gain difference between these two points are read from the diagram, Fig. 2.7. In this example, the gain

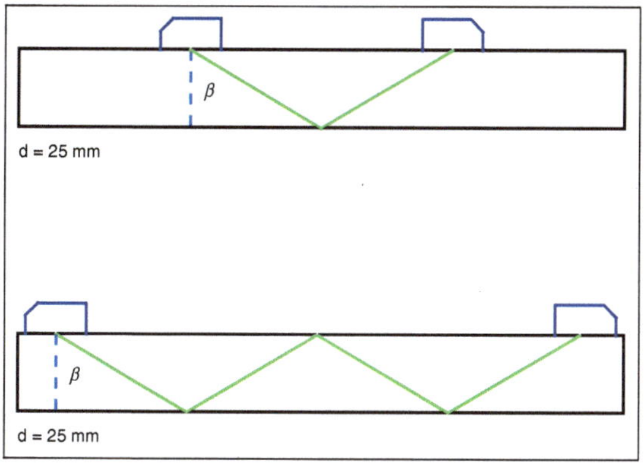

Fig. 2.6 V and W through transmission

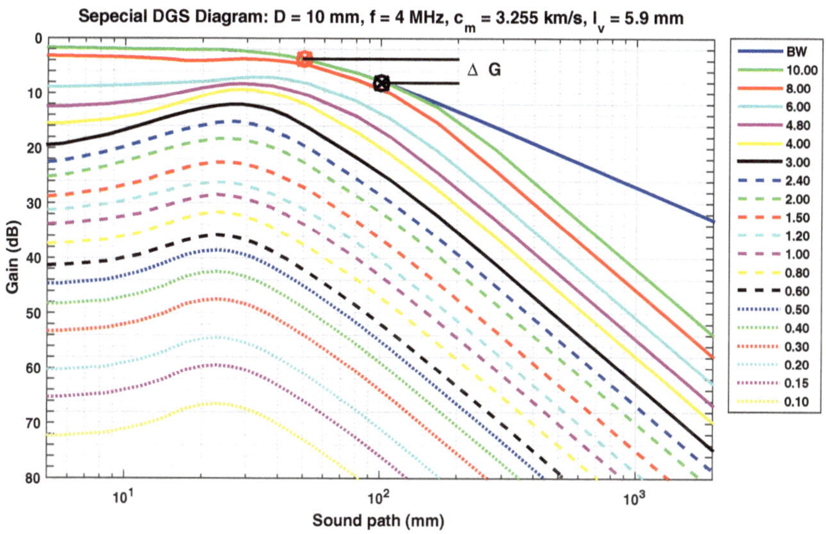

Fig. 2.7 Determination of the sound attenuation

difference read from the DGS diagram amounts to $\Delta G = 3.9\,\text{dB}$. Let the measured gain difference between V and W through transmission be 5 dB. This difference results from two influences, one is the difference based on the different sound paths read from the DGS diagram and the other one is the influence of the sound attenuation. That means, in the example, the influence of the sound attenuation having a sound path difference of 50 mm amounts to $\Delta G_k = 5\,\text{dB} - 3.9\,\text{dB} = 1.1\,\text{dB}$. Knowing this value enables the calculation of the sound attenuation κ_k in the calibration standard

$$\kappa_k = \frac{1.1}{2 \times 50}\,\frac{\text{dB}}{\text{mm}} = 0.011\,\frac{\text{dB}}{\text{mm}} = 11\,\frac{\text{dB}}{\text{m}} \tag{2.7}$$

The sound attenuation κ_m is determined accordingly. Let us assume that the result would be

$$\kappa_m = 15\,\frac{\text{dB}}{\text{m}}$$

In the next step, the transfer correction ΔV_t has to be derived. Therefore, the gain values and the sound paths of the V through transmissions on the calibration block and on the test specimen have to be known. The gain difference of these two measurements is due to several influences

- transfer correction
- different sound paths
- sound attenuation in the calibration block
- sound attenuation in the test specimen

To determine the transfer correction, three influences have to be eliminated from the measured gain difference. First, we need the sound path in the calibration block, which is already known from the example as 50 mm. Let us assume that the sound path for the through transmission on the test piece is 100 mm. With this the influences of the sound attenuations can be derived

$$\begin{aligned} V_k &= \frac{2 \times 50 \times 11}{1000}\,\text{dB} = 1.1\,\text{dB} \\ V_m &= \frac{2 \times 100 \times 15}{1000}\,\text{dB} = 3.0\,\text{dB} \end{aligned} \tag{2.8}$$

Let the gain value for setting the echo of the through transmission on the calibration standard to 80 % screen height be $G_k = 9.8\,\text{dB}$, and the gain setting for the measurement on the test piece be $G_m = 17.6\,\text{dB}$ accordingly. Since the DGS diagram does not take any sound attenuation into account the measured values have to be adapted accordingly. The measured gain values have to be corrected due to the sound attenuations. If no sound attenuation were active, the resulting echoes would have a larger amplitude; therefore, the gain settings have to be corrected by

$$\begin{aligned} &G_k - 1.1\,\text{dB} \\ &G_m - 3.0\,\text{dB} \end{aligned}$$

The difference of these two values results, using the assumed values, to 5.9 dB. The influence based on the different sound paths is 3.9 dB as already known from the DGS diagram for these sound paths. With this the transfer correction results in $\Delta V_t = 2\,\text{dB}$.

Now, all values needed for the DGS evaluation are known. Let the gain setting for 80 % screen height of the reference echo from the 25 mm arc of the calibration block K2 be $G_{K2} = 6\,\text{dB}$. An echo of a reflector found at a sound path of 45 mm requires a gain setting of $G_R = 36.6\,\text{dB}$ for the screen height of 80 %.

First, the measurement value of the **reference echo** is corrected for the use of the DGS method. If no sound attenuation would be existing, the echo would be higher than 80 %. The measured gain setting has to be reduced by the value of the sound attenuation $V_{\kappa\,K2}(25\,\text{mm})$

$$G_{K2} - V_{\kappa\,K2}(25\,\text{mm})$$

If the reference echo came from a plane back wall the echo would be lower by ΔV_{K2}, therefore, the gain value needs to be increased by ΔV_{K2}

$$G_{K2} - V_{\kappa\,K2}(25\,\text{mm}) + \Delta V_{K2} \tag{2.9}$$

Now, the correction of the measured gain setting for the **reflector** is performed accordingly for the sound attenuation $V_{\kappa\,m}$ in the test material. The measured gain setting is reduced by the value of the sound attenuation in the test material

$$G_R - V_{\kappa\,m}(45\,\text{mm})$$

Would the surface quality of test piece be as good as the surface of the calibration block, the resulting echo from the reflector would be larger by ΔV_t than 80 % screen height. The correction results accordingly

$$G_R - V_{\kappa\,m}(45\,\text{mm}) - \Delta V_t \tag{2.10}$$

Now, all corrections for the DGS evaluation are done. The difference of the two corrected values for the reflector echo, Eq. (2.10), and the corrected value for the reference echo, Eq. (2.9), can be derived

$$\Delta V = G_R - V_{\kappa\,m}(45\,\text{mm}) + V_{\kappa\,K2}(25\,\text{mm}) - G_{K2} - \Delta V_t - \Delta V_{K2} \tag{2.11}$$

The values for the sound attenuation have to be calculated

$$\begin{aligned} V_{\kappa\,K2}(25\,\text{mm}) &= \tfrac{2\times 25\times 11}{1000}\,\text{dB} = 0.55\,\text{dB} \\ V_{k\,m}(45\,\text{mm}) &= \tfrac{2\times 25\times 11}{1000}\,\text{dB} = 1.35\,\text{dB} \end{aligned}$$

The amplitude correction value ΔV_{K2} from the probe used for the example is zero. All values needed are now known and can be used in Eq. (2.11)

$$\Delta V = (36.6 - 1.35 + 0.55 - 6 - 2 - 0)\,\text{dB} = 27.8\,\text{dB} \tag{2.12}$$

With this gain difference, the DGS evaluation can be performed. For the reference echo, a point at a sound path of 25 mm is marked on the back wall curve of the DGS diagram. At $\Delta V = 27.8\,\text{dB}$ below this point a parallel line to the x-axis is drawn. At the intersection of this line with the sound path of 45 mm the equivalent reflector size (ERS) can be read from the DGS diagram; in the example, the result is $ERS = 1.2\,\text{mm}$ (Fig. 2.8).

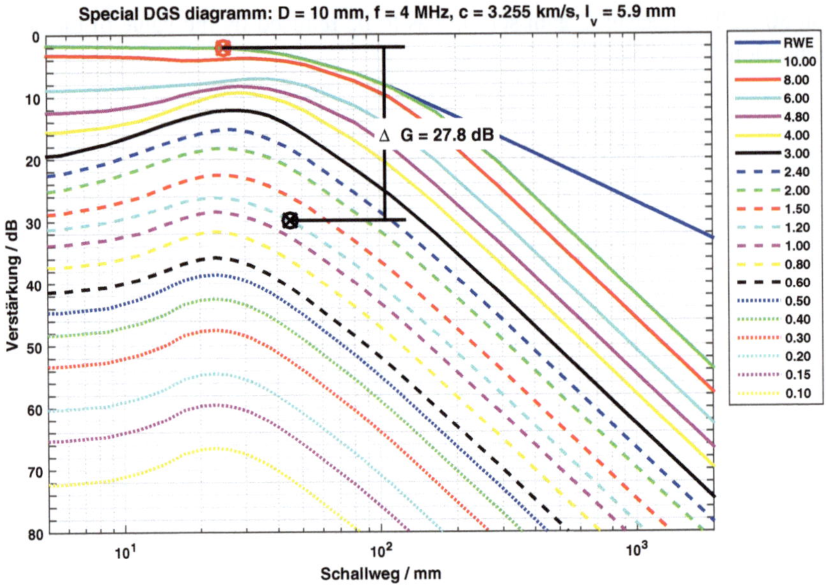

Fig. 2.8 DGS evaluation for a measurement using an angle beam probe

2.2.2.3 DGS and Ultrasonic Instruments

The DGS evaluation seems to by quite complex and sophisticated. But in earlier times, with analog ultrasonic instruments, DGS scales (design Kraukrämer) were used, Fig. 2.9. In today's digital ultrasonic instruments, DGS functionality is incorporated supporting the operator significantly. Already, in 1993, a patent *Flaw Detector incorporating DGS, US 5,511,425* was filed by Krautkrämer. In 1996, the patent was granted. Figure 2.10 shows the screen of a modern digital ultrasonic instrument with an incorporated DGS function. The curve for the selected equivalent reflector size is

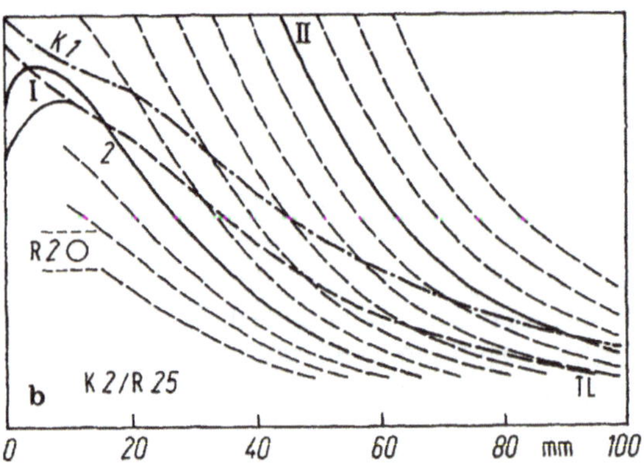

Fig. 2.9 DGS scale design Krautkrämer [1]

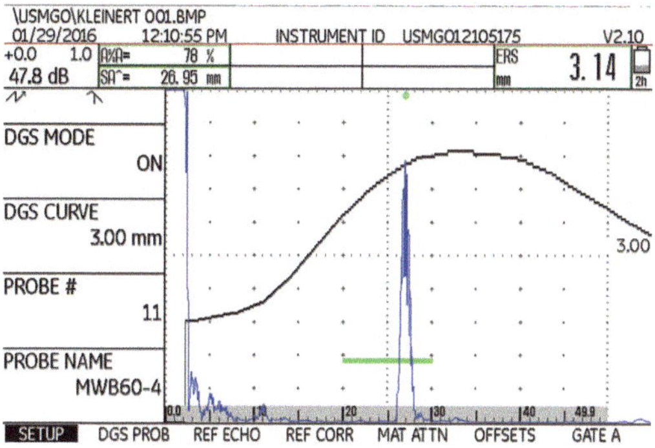

Fig. 2.10 Ultrasonic instrument with a DGS curve and DGS evaluation

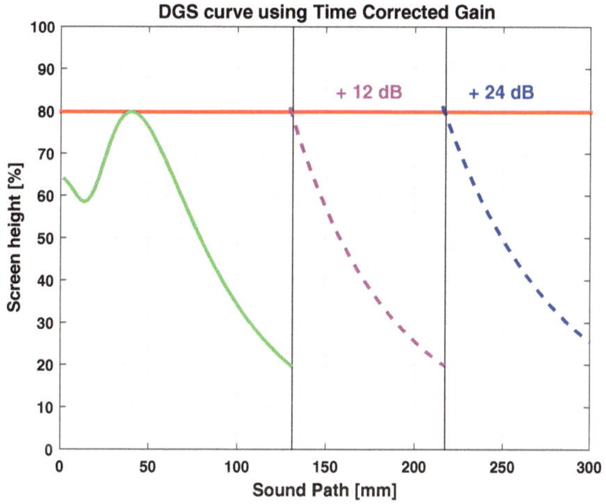

Fig. 2.11 Ultrasonic instrument with Time Corrected Gain according to a DGS curve

displayed on the screen. The actual ERS of the echo under evaluation can directly be read from the screen; in the figure, the value is $ERS = 3.14\,\text{mm}$. Alternatively, Time Corrected Gain (TCG) can be used. This function ensures that all echoes just reaching the DGS curve are set to 80 % screen height automatically (Fig. 2.11).

References

1. Krautkramer, J., Krautkramer, H.: Ultrasonic Testing of Materials. 4th Fully Revised Edition Translation of the 5th Revised German Edition. Springer, Berlin (1990)
2. Krautkrämer, J.: Fehlergrössenermittlung mit Ultraschall, Arch. Eisenhüttenwesen 30 (1959)
3. Ultrasonic testing - Sensitivity and range setting (ISO 16811:2012); German version EN ISO 16811:2014

Fig. 2.10 Ultrasonic [illegible]

Chapter 3
DGS Deviations Using Angle Beam Probes

Abstract Oversizing occurs using the DGS method with traditional angle beam probes. Most angle beam probes have rectangular transducers. In order to find an explanation of the deviations, the sound fields of straight beam probes with circular and rectangular transducers are analyzed and compared. Straight beam probes with circular transducers generate a rotationally symmetric sound field. Due to the effects at the interface between the probe wedge and the test piece, the generated sound field is far from rotationally symmetry-independent on whether a circular or a rectangular transducer is used in angle beam probes. Using the same DGS diagram for sizing with these different sound fields does not look promising. The observed issue of the deviations is not an issue from a single probe manufacturer, but a common problem with traditional angle beam probes.

For testing a new DGS software in an ultrasonic instrument, measurements were taken using special test blocks. These test blocks have plane back walls for 45°, 60°, and 70° angle beam probes. Into these back walls flat-bottomed holes with a diameter of 3 mm are drilled in different depths. Evaluating these measurements using the DGS method according to the state of the art [1, 2] leads to a surprise. In the range of a few near field lengths oversizing of the flat-bottomed holes (FBH) occurred, Fig. 3.1. In Fig. 3.1, the DGS evaluation is performed under the assumption that a sound attenuation does not have to be considered at a frequency of 2 MHz. Just to be sure the evaluation was repeated, this time with a sound attenuation of 10 dB/m, Fig. 3.2. But as can be seen from Fig. 3.2, this does not solve the issue with the DGS evaluation. The first assumption that something was wrong with the software had to be given up quite quickly. This was the starting point of evaluating the root causes for these deviations which finally led to the development of a new probe technology for angle beam probes.

W. Kleinert, *Defect Sizing Using Non-destructive Ultrasonic Testing*,
DOI 10.1007/978-3-319-32836-2_3

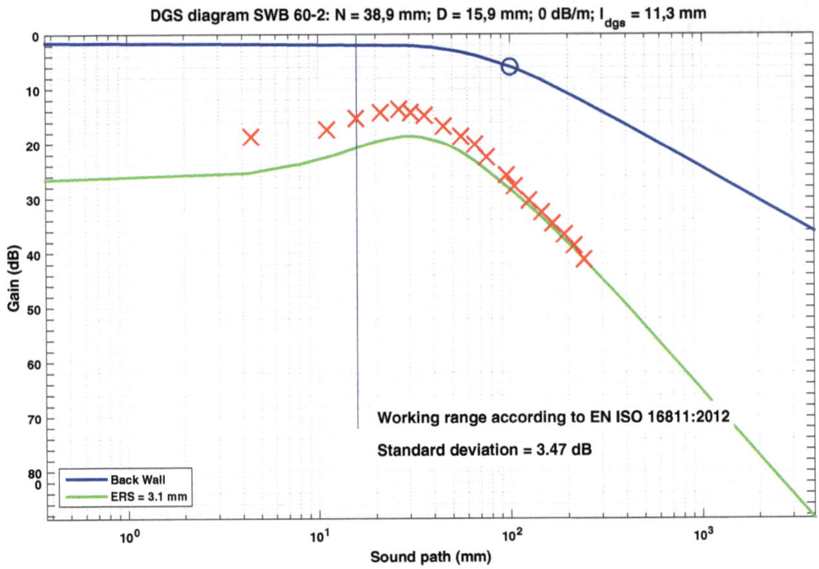

Fig. 3.1 Deviations in the DGS evaluation using a SWB 60-2

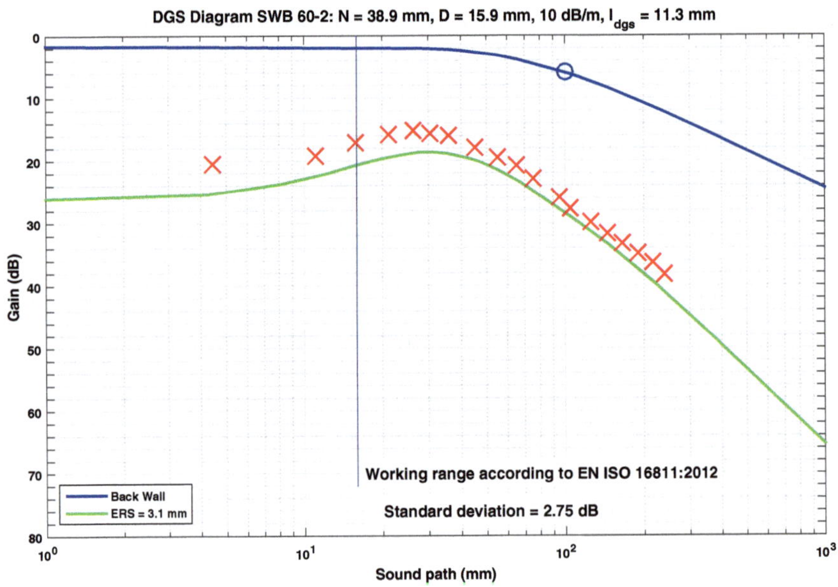

Fig. 3.2 DGS evaluation of measurements taken with a SWB 60-2 considering a sound attenuation of 10 dB/m

3.1 Sound Fields

As already mentioned, the DGS method was developed for straight beam probes with flat circular transducers. This method is used as well to evaluate measurements taken with an angle beam probe having a rectangular transducer. Straight beam probes with flat circular transducers generate a rotationally symmetric sound field. In the first step of discovering the root cause for the oversizing with an angle beam probe, the sound field of such a probe was simulated with the commercially available software, CIVA. The result is shown in Fig. 3.3. It is obvious that the sound field of this angle beam probe is definitely not rotationally symmetric at all. This finding led to further evaluations of sound fields. First, the sound field long sections through the acoustic axis of a rectangular transducer ($9 \times 8\,\text{mm}^2$) were calculated for straight beam insonification. All sound field calculations are based on continuous sound (mono

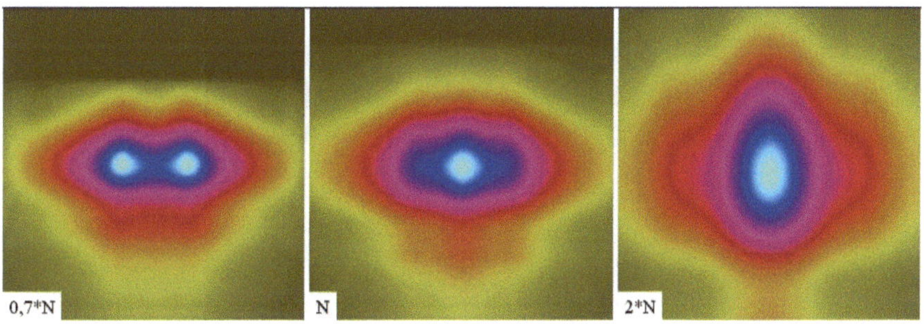

Fig. 3.3 Sound field cross sections of an angle beam probe with a $8 \times 9\,\text{mm}^2$ transducer at 0.7, 1, and 2 near field lengths

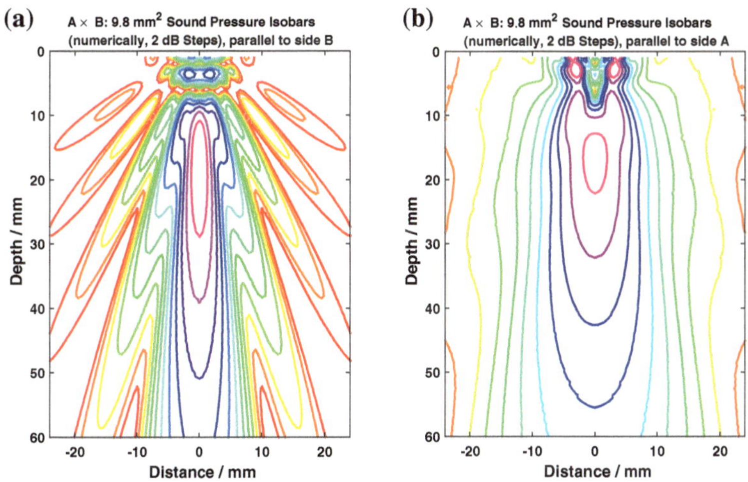

Fig. 3.4 Long sections through the sound field of a 8×9 mm transducer through the acoustic axis: **a** parallel to the longer side. **b** Parallel to the shorter side

frequent). Figure 3.4 shows these long sections of the sound field as isobars in 2 dB steps. No rotational symmetry exists. In the next step, the cross sections of straight beam probes were calculated. The result of a circular transducer was compared to the result of a rectangular transducer. The cross sections were calculated at depths of 0.5, 1, and 3 near field lengths N. Comparing the cross sections of a rectangular transducer to the cross sections of a circular transducer shows how significant the differences are (Fig. 3.5). That the measurements with both transducer shapes can be evaluated correctly with the same DGS diagram is not very plausible. The sound field sections have been calculated for straight beam insonification. Using angle beam probes with rectangular transducers will influence the sound field additionally because of the refraction and the mode conversion at the interface between probe wedge and test piece.

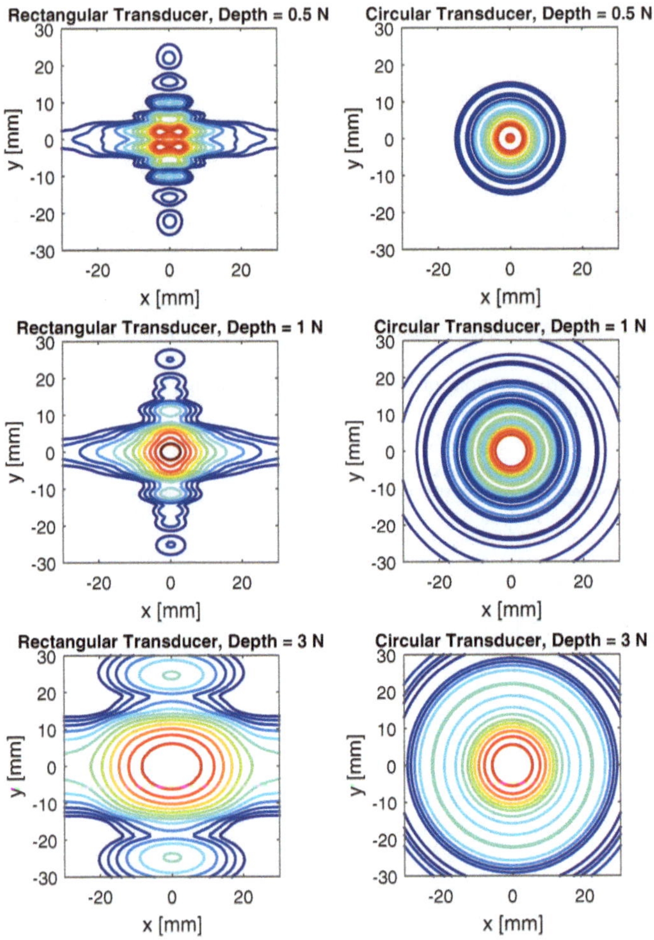

Fig. 3.5 Cross sections perpendicular to the acoustic axis in the depth of 0.5, 1 and 3 near field lengths

3.2 A Manufacturer-Independent Issue

The findings of the deviations using the DGS method with angle beam probes were first reported at the 10th European Conference on Non-Destructive Testing (ECNDT), in Moscow 2010 [3]. In the same year, Olympus published after the ECNDT an article with the title *DGS Sizing Diagram with Single Element and Phased Array Angle Beam Probe* [4] published by NDT.net. They used a different approach to the observed deviations using the DGS method with traditional angle beam probes. They compared the DGS diagrams of 2 and 4 MHz probe with CIVA simulations of the echo amplitude from flat-bottomed holes. This approach confirmed the observed deviations, Fig. 3.6. The caption of Fig. 3.6 in the Olympus article is

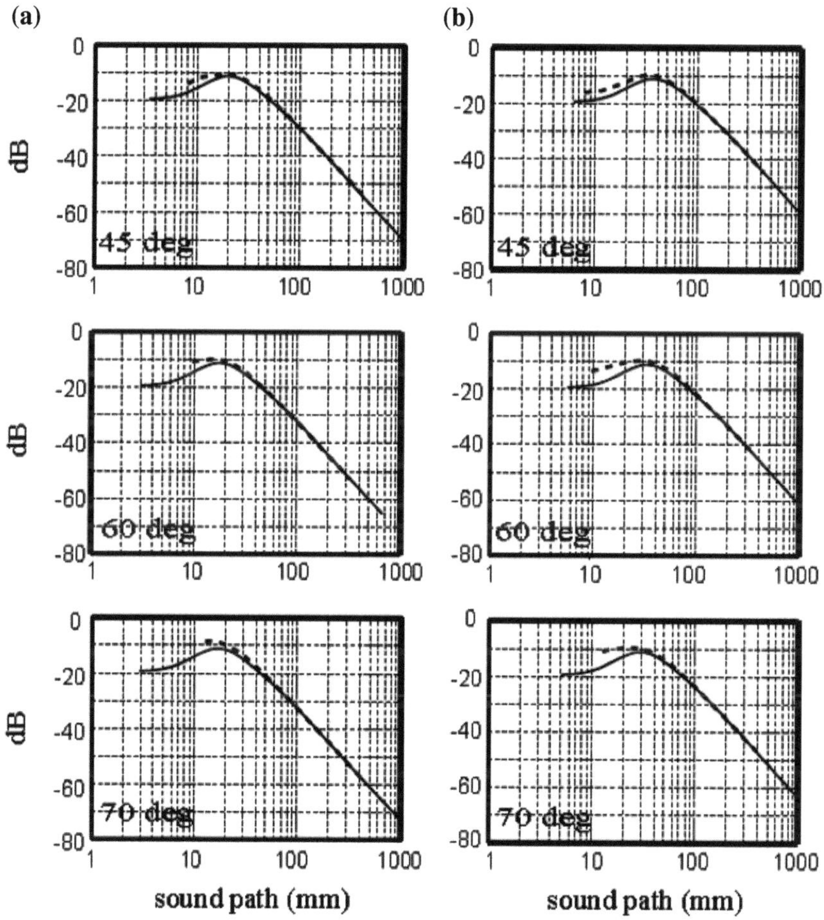

Fig. 3.6 Excerpt from an Olympus article published at ndt.net [4]

Comparison of universal DGS curves (solid line) with CIVA model curves (dashed line). The target is 0.125 in. diameter FBH. a: AM2R8x9-45, -60, -70. b: AM4R8x9-45, -60, -70.

The observed deviations occur while using the DGS method with traditional angle beam probes and are not an issue of a single manufacturer of angle beam probes. This issue is based on the fact that the DGS method was developed for straight beam probes generating a rotationally symmetric sound field. The sound fields of traditional angle beam probes are significantly different resulting in the observed deviations.

3.3 The Beginning of a New Probe Technology

It was understood that the deviations have a physical root cause in the not existing rotational symmetry of traditional angle beam probes. This is independent of whether the angle beam probes are built with a rectangular transducer or with a circular transducer. The symmetry will be further disturbed based on the refraction, the mode conversion, and the angle-dependent phase shift at the interface between the wedge of the probe and the test piece.

The question came up: how an angle beam probe can be designed that generates a rotationally symmetric sound field? This thought was the beginning of a much better understanding of the DGS method including the bandwidth dependency. This new probe technology improved not only the accuracy of DGS evaluations with angle beam probes but enabled as well the possibility to calculate Distance Amplitude Correction curves (DAC) [5].

The assignment of the task is shown in Fig. 3.7.

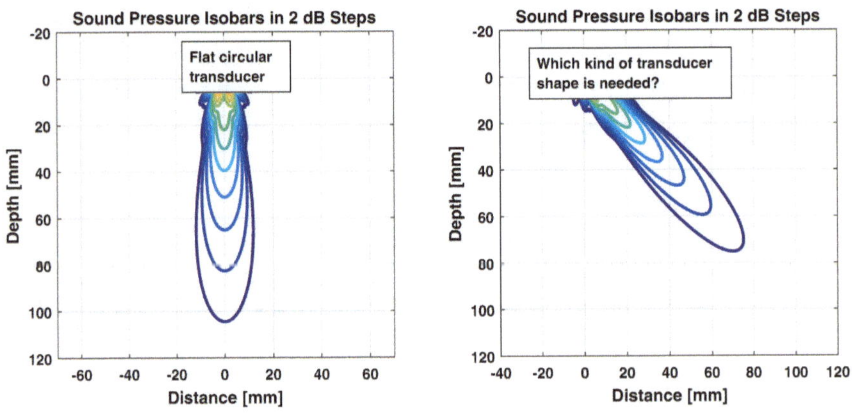

Fig. 3.7 The task to be solved

References

1. Krautkramer, J., Krautkramer, H.: Ultrasonic Testing of Materials. 4th Fully Revised Edition Translation of the 5th Revised German Edition, Springer, Berlin (1990)
2. Krautkrämer J.: Fehlergrössenermittlung mit Ultraschall, Arch. Eisenhüttenwesen 30 (1959)
3. Kleinert W., Oberdörfer Y., Splitt G.: The Ideal Angle Beam Probe for DGS Evaluation, 10th ECNDT, Moskau (2010). http://www.ndt.net/article/ecndt2010/reports/1_03_64.pdf
4. Yan X., Zhang J., Kass D., Langlois P.: DGS Sizing Diagram with Single Element and Phased Array Angle Beam Probe, Olympus, Waltham, MA, ndt.net (2010). http://www.ndt.net/article/ndtnet/2010/14_Yan.pdf
5. Kleinert, W., Oberdörfer, Y.: Calculated Bandwidth Dependent DGS and DAC Curves for Phased Array Sizing. In: Proceedings of ECNDT Prag (2014). http://www.ndt.net/events/ECNDT2014/app/content/Paper/165_Kleinert.pdf

Chapter 4
The New Probe Technology, Single Element Probes

Abstract The design principle of the new probe technology is introduced. The calculation of a point cloud defining the transducer shape is discussed in detail. Necessary adaptations are taken into account. The new angle beam probes generate rotationally symmetric sound fields as expected. The DGS accuracy in the working range defined in the EN ISO 16811:2011 is significantly better than the accuracy achieved with traditional angle beam probes.

The new probe technology was developed at GE Sensing & Inspection Technologies GmbH in Hürth, Germany. In the following, the new technology is referred to as *true*DGS® technology.[1] The basic starting point is the question, how to design an angle beam probe that generates a rotationally symmetric sound field, Fig. 3.7.

4.1 Design Principle

The task is to transfer the sound field of a straight beam probe for transverse waves with a circular transducer to an angle beam probe. The goal is to calculate the transducer shape of an angle beam probe in such a way that a rotationally symmetric sound field is generated.

The basic idea is illustrated in Fig. 4.1. The left side shows the near field of a straight beam probe with a circular transducer having a diameter D calculated for shear waves. The right side represents the near field of the angle beam probe under construction. The goal is to design an angle beam probe with a given angle of incidence β and a predefined delay v_w. First, the time of flight from the straight beam transducer to the end of its near field is calculated. From this time of flight, the time in the delay of v_w in the wedge of the angle beam probe under construction is subtracted. The remaining time defines the end of the near field of the angle beam probe under construction (point T) in the right side of Fig. 4.1 considering the predefined angle of

[1] true*DGS*® is a registered trademark of GE Sensing & Inspection Technologies GmbH.

W. Kleinert, *Defect Sizing Using Non-destructive Ultrasonic Testing*,
DOI 10.1007/978-3-319-32836-2_4

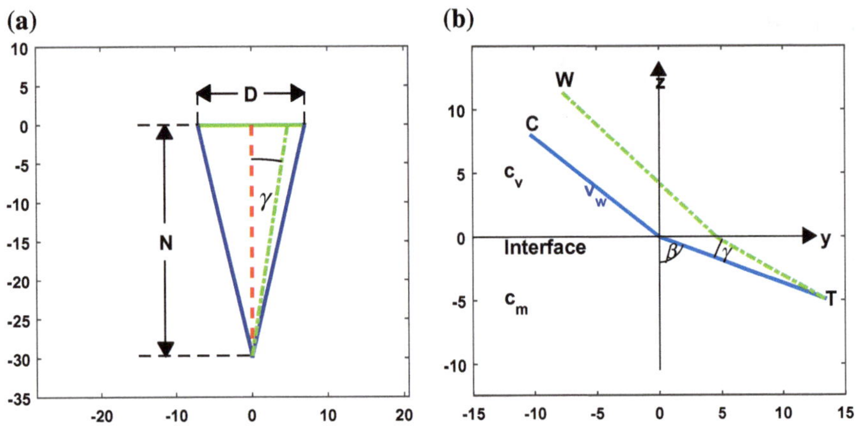

Fig. 4.1 Basic idea for the new probe technology: **a** Straight beam probe used as base of the design. **b** Angle beam probe under construction

incidence β and the sound velocity c_m in the test material. By this, the first point on the transducer of the angle beam probe is derived: point C in the right part of Fig. 4.1. For each angle γ, the time of flight t_γ is calculated for the straight beam probe. This time of flight is transferred to the angle beam probe using the same angle γ. The time of flight from the end of the near field (point T) to the interface between probe and test piece is calculated and subtracted from the time t_γ. The remaining time defines the point W in the right part of Fig. 4.1 under consideration of Snell's Law and the sound velocity c_v in the wedge of the probe. With point W, another point on the transducer under construction is defined. Repeating this process for all possible angles γ in the plane of projection and in the plane perpendicular to the plane of projection results in a point cloud defining the transducer shape (Fig. 4.2). But this is not yet the final solution. Additional effects have to be taken into account

- Angle-dependent phase shift
- The intended angle of incidence is changed due to the different phase shifts of the rim beams. Therefore, a correction angle has to be applied.
- The design of a true*DGS*® transducer based on a circular straight beam transducer with diameter D is not area invariant. For the DGS evaluation, a correction factor must to be applied.

The complete calculation method including these required adaptations is described in Sect. 4.2 in detail [5, 6].

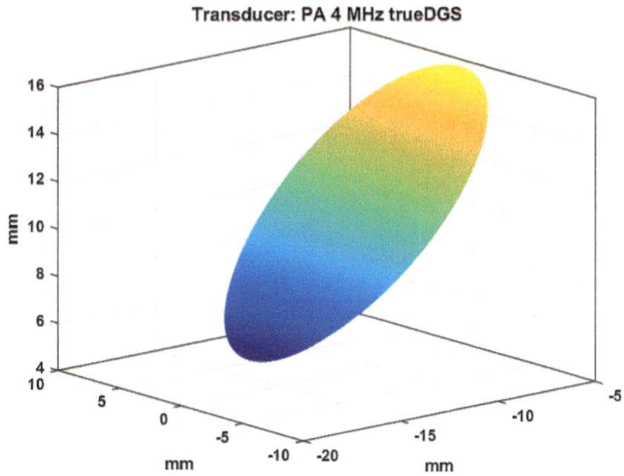

Fig. 4.2 Example of a transducer of a *true*DGS® probe

4.2 Calculation Method

All calculations are based on the coordinate system shown in Fig. 4.3. The coupling surface is defined by the x-y plane. Point T describes the end of the near field of the angle beam probe. Point D (puncture point) is the intersection of the sound beam with the coupling surface. Point W is one point within the point cloud defining the transducer shape.

The vector **z** connects the end of the near field T (target point) with the origin of the coordinate system. Vector **d** describes a single sound beam in the test material starting at the target point and ending at the intersection with the coupling surface. The resulting sound beam in the wedge of the probe is represented by vector **v**.

Let the target point T have the coordinates $(0; y_T; z_T)$. Let the puncture point D on the coupling surface have the coordinates $(\overline{x}; \overline{y})$ defining vector $\overline{\boldsymbol{a}}$. The vectors can now be defined as

$$\mathbf{z} = \begin{pmatrix} 0 \\ -y_T \\ -z_T \end{pmatrix} \tag{4.1}$$

The vector **d** results from the sum of the vectors **a** and **z**, refer to Fig. 4.3:

$$\mathbf{d} = \begin{pmatrix} \overline{x} \\ \overline{y} - y_T \\ -z_T \end{pmatrix} \tag{4.2}$$

Let vector **b** be the vector from the origin of the coordinate system to the transducer point W

$$\mathbf{b} = \begin{pmatrix} x_w \\ y_w \\ z_w \end{pmatrix} \tag{4.3}$$

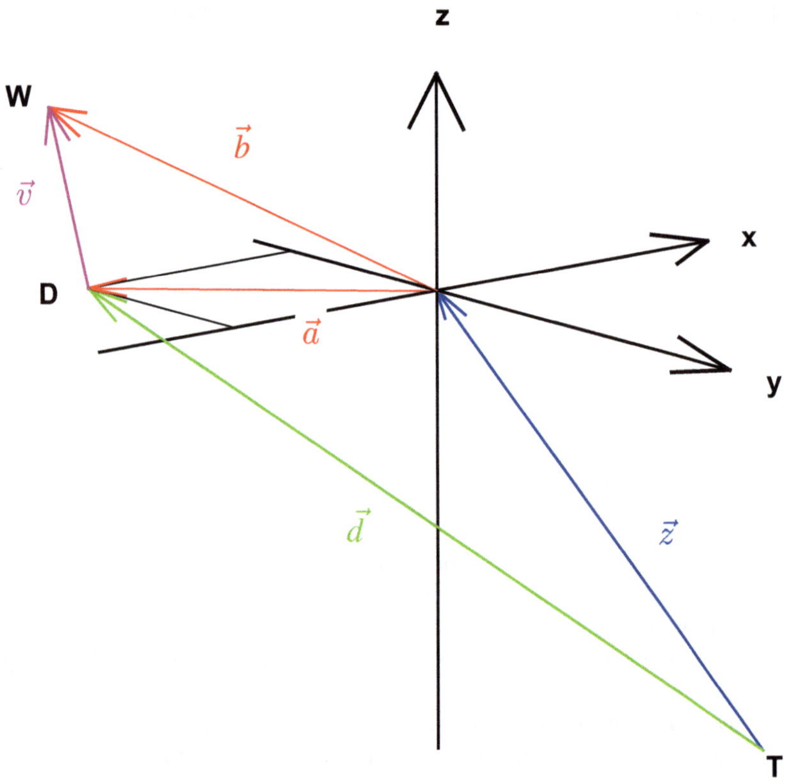

Fig. 4.3 Coordinate system used for all calculations

The difference of the vectors **b** and vector **a** defines vector **v**:

$$\mathbf{v} = \begin{pmatrix} x_w - \overline{x} \\ y_w - \overline{y} \\ z_w \end{pmatrix} \tag{4.4}$$

4.2.1 *The Fastest Path*

The fastest path between point W on the transducer to the target point T (end of the near field of the angle beam probe) is examined. For this calculation, the following parameters are used:

- d_v: distance in the wedge of the probe
- d_m: distance in the test material
- t: total time of flight

The distance d_v in the wedge of the probe is given by $d_v = |\mathbf{v}|$

$$d_v = \sqrt{(x_w - \overline{x})^2 + (y_w - \overline{y})^2 + z_w^2} \tag{4.5}$$

Accordingly, the distance in the test material is given by $d_m = |\mathbf{d}|$

$$d_m = \sqrt{\overline{x}^2 + (\overline{y} - y_T)^2 + z_T^2} \tag{4.6}$$

With this the total time of flight t results in

$$t = \frac{1}{c_v}\sqrt{(x_w - \overline{x})^2 + (y_w - \overline{y})^2 + z_w^2} + \frac{1}{c_m}\sqrt{\overline{x}^2 + (\overline{y} - y_T)^2 + z_T^2} \tag{4.7}$$

with:

- c_v: sound velocity in the wedge of the probe
- c_m: sound velocity in the test material

The partial derivatives of the time of flight t, Eq. (4.7), with respect to $\overline{x}$ and $\overline{y}$ are calculated and set to zero to determine the minimum time of flight

$$\begin{aligned} \frac{\partial t}{\partial \overline{x}} &= \frac{1}{c_v}\frac{x_w - \overline{x}}{\sqrt{(x_w-\overline{x})^2+(y_w-\overline{y})^2+z_w^2}} - \frac{1}{c_m}\frac{\overline{x}}{\sqrt{\overline{x}^2 + (\overline{y} - y_T)^2 + z_T^2}} = 0 \\ \frac{\partial t}{\partial \overline{y}} &= \frac{1}{c_v}\frac{y_w - \overline{y}}{\sqrt{(x_w-\overline{x})^2+(y_w-\overline{y})^2+z_w^2}} - \frac{1}{c_m}\frac{\overline{y} - y_T}{\sqrt{\overline{x}^2 + (\overline{y} - y_T)^2 + z_T^2}} = 0 \end{aligned} \tag{4.8}$$

For further examination, the unit vectors $\mathbf{e_x}$ and $\mathbf{e_y}$ are introduced

$$\begin{aligned} \mathbf{e_x} &= \begin{pmatrix} 1 \\ 0 \\ 0 \end{pmatrix} \\ \mathbf{e_y} &= \begin{pmatrix} 0 \\ 1 \\ 0 \end{pmatrix} \end{aligned} \tag{4.9}$$

Using these unit vectors, the series of Eq. (4.8) can be written more easily. Initially, the first equation of the series is considered

$$\begin{aligned} \frac{1}{c_v}\frac{\mathbf{v}\cdot\mathbf{e_x}}{|\mathbf{v}|\,|\mathbf{e_x}|} &= \frac{1}{c_m}\frac{\mathbf{d}\cdot\mathbf{e_x}}{|\mathbf{d}|\,|\mathbf{e_x}|} \\ \frac{1}{c_v}\cos\,\alpha_x &= \frac{1}{c_m}\cos\,\beta_x \\ \Rightarrow \frac{\cos\,\alpha_x}{\cos\,\beta_x} &= \frac{c_v}{c_m} \end{aligned} \tag{4.10}$$

with:

- α_x: angle between the sound beam and the x direction in the wedge
- β_x: angle between the sound beam and the x direction in the test material

The second equation of the series (4.8) is dealt with accordingly resulting in:

$$\frac{\cos\ \alpha_y}{\cos\ \beta_y} = \frac{c_v}{c_m} \tag{4.11}$$

- α_y: angle between the sound beam and the y-direction in the wedge
- β_y: angle between the sound beam and the y-direction in the test material

This is nothing but a three dimensional version of Snell's Law, but it has to be noted that here the angles between the sound beams and the axes are used and not the angles to the perpendicular lines to the axes as usual.

4.2.2 *Included Angle*

For a preset point D on the coupling surface, refer to Fig. 4.3, with the coordinates $(\overline{x}; \overline{y}; 0)$ the angle γ between the central beam and the sound beam under consideration is derived. This angle γ is included by the vectors **d** and **z**, hence

$$\cos\ \gamma = \frac{\mathbf{d} \cdot \mathbf{z}}{|\mathbf{d}|\ |\mathbf{z}|} \tag{4.12}$$

4.2.3 *Time of Flight*

The time of flight in the test material follows from:

$$t_m = \frac{|\mathbf{d}|}{c_m} \Rightarrow t_m = \frac{1}{c_m} \sqrt{\overline{x}^2 + (\overline{y} - y_T)^2 + z_T^2} \tag{4.13}$$

The time of flight $t(\gamma)$ for the straight beam probe results in, refer Fig. 4.1:

$$t(\gamma) = \frac{N}{c_m\ \cos\ \gamma} \tag{4.14}$$

with: N near field length of the predefined straight beam probe.

Hence, the distance d_v in the wedge for the sound beam under construction can be derived

$$t_v = t(\gamma) - t_m, \; d_v = t_v \, c_v \tag{4.15}$$

The distance d_v equals $|\mathbf{v}|$. With this, the time of flight in the wedge is given by

$$t_v = \frac{|\mathbf{v}|}{c_v} \Rightarrow t_v = \frac{1}{c_v} \sqrt{(x_w - \overline{x})^2 + (y_w - \overline{y})^2 + z_w^2} \tag{4.16}$$

4.2.4 Angle in the Test Material

The angles in the test material between the sound beam under construction and the axes are identified as

- β_x: angle between the sound beam under construction and the x-axis
- β_y: angle between the sound beam under construction and the y-axis

These angles can be calculated using

$$\cos\, \beta_x = \frac{\mathbf{d} \cdot \mathbf{e_x}}{|\mathbf{d}| \; |\mathbf{e_x}|} = \frac{\overline{x}}{\sqrt{\overline{x}^2 + (\overline{y} - y_t)^2 + z_T^2}} \tag{4.17}$$

and:

$$\cos\, \beta_y = \frac{\mathbf{d} \cdot \mathbf{e_y}}{|\mathbf{d}| \; |\mathbf{e_y}|} = \frac{\overline{y} - y_T}{\sqrt{\overline{x}^2 + (\overline{y} - y_t)^2 + z_T^2}} \tag{4.18}$$

4.2.5 Angles in the Wedge of the Probe

From Eqs. (4.10) and (4.11) follows:

$$\begin{aligned} \cos\, \alpha_x &= \frac{c_v}{c_m} \cos\, \beta_x \\ \cos\, \alpha_y &= \frac{c_v}{c_m} \cos\, \beta_y \end{aligned} \tag{4.19}$$

In addition, the angles α_x and α_y can be calculated using the following dot products:

$$\cos\, \alpha_x = \frac{\mathbf{v} \cdot \mathbf{e_x}}{|\mathbf{v}| \; |\mathbf{e_x}|} = \frac{x_w - \overline{x}}{\sqrt{(x_w - \overline{x})^2 + (y_w - \overline{y})^2 + z_w^2}}, \tag{4.20}$$

respectively

$$\cos\ \alpha_y = \frac{\mathbf{v} \cdot \mathbf{e_y}}{|\mathbf{v}|\ |\mathbf{e_y}|} = \frac{y_w - \overline{y}}{\sqrt{(x_w - \overline{x})^2 + (y_w - \overline{y})^2 + z_w^2}}. \tag{4.21}$$

Using Eqs. (4.10) and (4.11) the angles β_x and β_y can be derived

$$\cos\ \beta_x = \frac{c_m}{c_v}\ \frac{x_w - \overline{x}}{\sqrt{(y_w - \overline{y})^2 + (y_w - \overline{y})^2 + z_w^2}} \tag{4.22}$$

$$\cos\ \beta_y = \frac{c_m}{c_v}\ \frac{y_w - \overline{y}}{\sqrt{(x_w - \overline{x})^2 + (y_w - \overline{y})^2 + z_w^2}}$$

4.2.6 *Transducer Coordinates*

Equation (4.16) can be converted to

$$t_v\ c_v = \sqrt{(x_w - \overline{x})^2 + (y_w - \overline{y})^2 + z_w^2} \tag{4.23}$$

In Eq. (4.22), the square root in the denominator is replaced according to Eq. (4.23) resulting in

$$\cos\ \beta_x = \frac{c_m}{c_v}\ \frac{x_w - \overline{x}}{t_v\ c_v} \tag{4.24}$$

$$\Rightarrow x_w = \frac{c_v^2\ t_v\ \cos\ \beta_x}{c_m} + \overline{x}$$

respectively:

$$\cos\ \beta_y = \frac{c_m}{c_v}\ \frac{y_w - \overline{y}}{t_v\ c_v} \tag{4.25}$$

$$\Rightarrow y_w = \frac{c_v^2\ t_v\ \cos\ \beta_y}{c_m} + \overline{y}$$

Now, the coordinates x_w and y_w can be calculated. Using Eqs. (4.24) and (4.25) in Eq. (4.23) results in the third coordinate z_w

$$z_w = \sqrt{t_v^2\ c_v^2 - (x_w - \overline{x})^2 - (y_w - \overline{y})^2} \tag{4.26}$$

The point cloud defining the transducer shape is calculated using the Eqs. (4.24)–(4.26) by varying the variables $\overline{x}$ and $\overline{y}$. By doing this it has to be ensured that the condition $\gamma \leq \gamma_{max}$ is fulfilled. The angle γ_{max} can be easily determined, Fig. 4.1

$$\gamma_{max} = \arctan \frac{D}{2N} \tag{4.27}$$

with:

- D: diameter of the predefined straight beam probe
- N: near field length of the predefined straight beam probe

Note: For the final result of the transducer shape, some adaptations are required which will be discussed in detail in Sect. 4.3.

4.2.7 Calculation Summary

With all these formulas, the overview of the calculation method can get lost quite quickly. Therefore, in the following, the different steps are listed without the use of any formula:

1. The frequency and the diameter of the straight beam probe as basis of the construction are chosen.
2. For this straight beam probe, the near field length N for shear waves is calculated and the angle γ_{max} is derived.
3. The angle of incidence β and the delay length v_w for the angle beam probe under construction are determined.
4. The time of flight t from the transducer to the end of the near field of the straight beam probe is derived.
5. From this time of flight t the time of flight in the delay length v_w is subtracted and the remaining time is used to calculate the coordinates of the near field end (point T) under consideration of Snell's Law, Fig. 4.1.
6. A point D with the coordinates $(\overline{x}; \overline{y}; 0)$ on the coupling surface is chosen.
7. The angle γ between the central axis and the vector from point T to point D on the coupling surface is calculated, Eq. (4.14).
8. The fulfillment of the condition $\gamma \leq \gamma_{max}$ is checked, Eq. (4.27).
9. For the angle γ, the time of flight $t(\gamma)$ for the straight beam probe is calculated, refer Fig. 4.1 and Eq. (4.12).
10. The length of the vector from point T (Fig. 4.1) to the point D selected on the coupling surface is derived.
11. Based on this length, the time of flight t_m in the test material is calculated.

12. The value of t_m is subtracted from the total time of flight $t(\gamma)$. Hence, the time of flight t_v in the wedge of the probe is known as well.
13. The angles β_x and β_y in the test material are calculated; Eqs. (4.17) and (4.18).
14. Now, the transducer coordinates x_w, y_w, and z_w can be calculated using Eqs. (4.24)–(4.26).

To calculate the complete point cloud of the transducer, the coordinates $\overline{x}$ and $\overline{y}$ have to be chosen accordingly and the steps 7 to 14 have to be repeated for each pair of $\overline{x}$ and $\overline{y}$.

4.3 Necessary Adaptations

Three necessary adaptations have to be implemented

- angle-dependent phase shift
- corrected angle of incidence
- area correction

4.3.1 Phase Shift

An angle-dependent phase shift originates at the interface between probe wedge and test material [1–3]. This phase shift exists at both directions of the sound traveling back and forth. Figure 4.4 shows the phase shifts for both directions. When using the DGS method the echo field has to be applied. Therefore, the sum of the phase shifts for the directions back and forth has to be taken into account, Fig. 4.5.

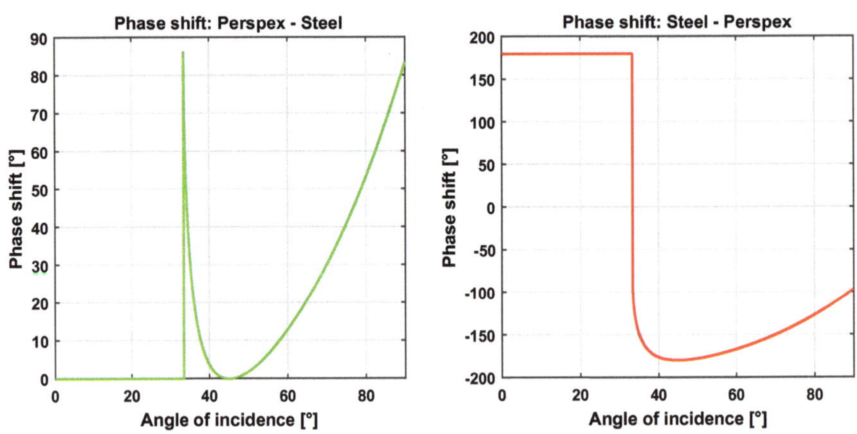

Fig. 4.4 Phase shifts at the interface between probe wedge and test material

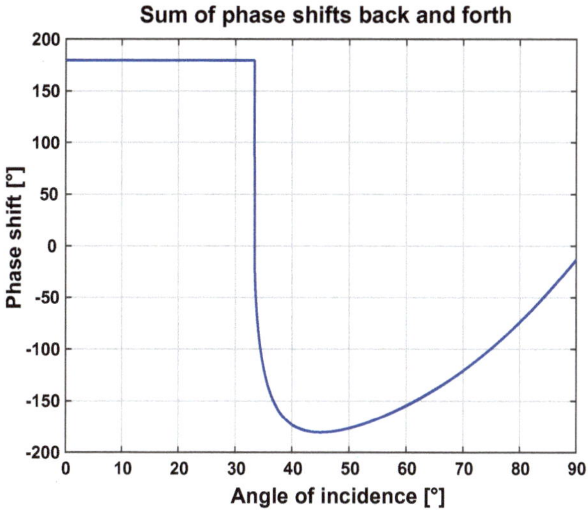

Fig. 4.5 Sum of the phase shifts for both directions back and forth

The formulas for calculating the phase shifts are published in the Kraukträmer book [1]. For easier reading, these formulas are included here. In the Krautkrämer book, only the real part of the formulas are considered. For calculating the phase shifts, the imaginary part has to be utilized as well.

The following abbreviations are used in the formulas:

- N: abbreviation for the denominator
- index 1: first material, here test material
- index 2: second material, here material of the probe wedge
- index t: identifies the transverse wave
- index l: identifies the longitudinal wave
- c: identifies the particular sound velocity
- ρ: identifies the particular mass density

According to the Krautkrämer book, for the incident longitudinal wave in material 1 applies

$$N = 2\ \cot\ \alpha_{1t} + \frac{\cos^2\ 2\alpha_{1t}}{2\ \sin^4\ \alpha_{1t}}\ \tan\ \alpha_{1l} + \frac{2\ \rho_2\ c_{2t}^4}{\rho_1\ c_{1t}^4}\ \cot\ \alpha_{2t} + \frac{\rho_2\ \cos^2\ 2\alpha_{2t}}{2\ \rho_1\ \sin^4\ \alpha_{1t}}\ \tan\ \alpha_{2l} \tag{4.28}$$

Since the incident longitudinal wave in material 2 has to be examined, the indices in Eq. (4.28) have to be inverted. This is valid as well for the following equation describing the transmission coefficient for the transition from the longitudinal wave to the transverse wave

$$D_{tl} = \frac{2\ \rho_2\ c_{2t}^2}{\rho_1\ c_{1t}^2}\ \frac{\cos\ 2\alpha_{1t}}{N\ \sin^2\ \alpha 1t} \tag{4.29}$$

With the inverted indices the direction from the transducer to the reflector is described. The transmission coefficient from the transverse wave in steel into the longitudinal wave in the wedge is given by (using inverted indices in Eq. (4.28))

$$D_{lt} = -\ \frac{4\ \rho_2\ c_{2l}^2}{\rho_1\ c_{1t}^2}\ \frac{\cos\ 2\alpha_{2t}}{N\ \sin\ 2\alpha_{2l}}\ \cot\ \alpha_{1t} \tag{4.30}$$

The phase shift is given by the ratio of the imaginary part and the real part. The resulting phase shifts by angle of incidence are shown in Fig. 4.4.

Figure 4.5 shows the sum of the phase shifts for both directions back and forth. In the following, the angle of incidence β_s of each individual sound beam in the sound field has to be considered for the angle-dependent phase shift. Let p_s be the phase shift in degree, β_s the particular angle of incidence and λ the wave length in the wedge material. The required adaptation to the calculated *true*DGS® transducer shape is realized by prolonging the calculated distances d_v in the wedge by v_s

$$v_s = -\ \frac{p_s\,(\beta_s)}{360°}\ \lambda \tag{4.31}$$

Hence, the calculation of the particular delay d_v in Eq. (4.15) needs to be changed to

$$d_v = t_v\ c_v + v_s \tag{4.32}$$

In Fig. 4.6, the exaggerated prolongations of the calculated distances d_v in the wedge are shown as dotted lines.

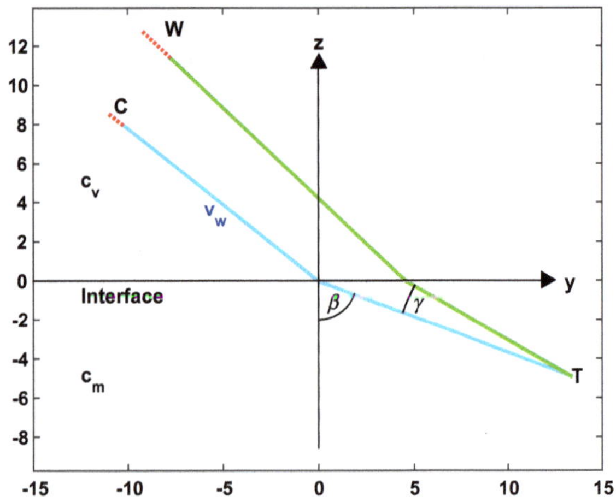

Fig. 4.6 Applied phase correction

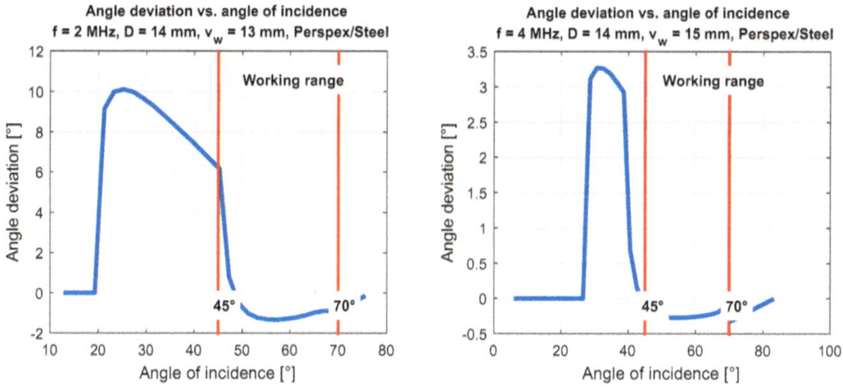

Fig. 4.7 Angle deviation in dependence on the angle of incidence

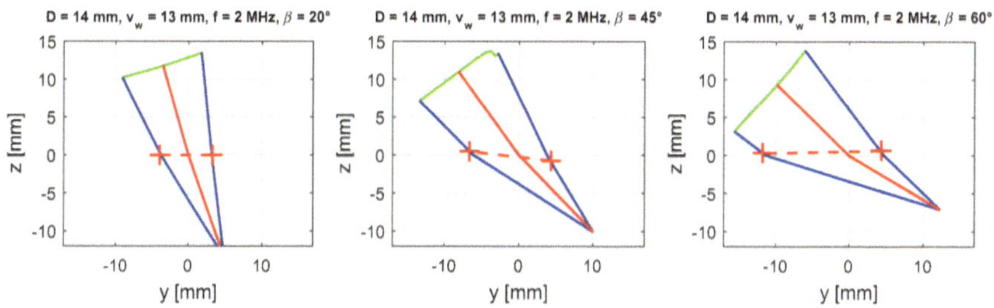

Fig. 4.8 Correction of the angle of incidence

4.3.2 *Corrected Angle of Incidence*

Based on the different phase shifts of the rim beams traveling from the wedge into the test material, the intended angle of incidence is changed, thus the sound field is tilted. Therefore, a correction of the angle of incidence has to be applied (Fig. 4.7). This correction value depends on the probe parameters and angle of incidence chosen. In the easiest case, the correction value is zero; refer to the left graph of Fig. 4.8.

The calculated tilting angle cannot be added to the indented angle of incidence, respectively, subtracted from the angle because with this new angle of incidence again tilting takes place. This problem can be solved numerically up to a predefined accuracy.

The values of the correction angles depend on many probe parameters such as transducer size, delay length, frequency, etc. The values for the correction angle increase significantly at incidence angles close to or below the point of non-differentiability of the phase shift graph.

4.3.3 *Area Correction*

The calculation method for *true*DGS® is not area invariant. The transducer area of the straight beam probe used for the calculation of the *true*DGS® probe is different to the area of the resulting transducer.

This fact has a few consequences

- When deriving a special DGS diagram from the general DGS diagram, the diameter of the straight beam probe used for the design cannot be multiplied with the size indication *G* in the general DGS diagram.
- The diameter D_{dgs} of a circular transducer having the same area as the *true*DGS® transducer has to be calculated and multiplied with the size indication.
- But for deriving the special DGS diagram, the near field length of the straight beam probe used as basis for the design of the angle beam probe has to be applied.

- Hence, the known formula for calculating the near field length, Eq. (2.1), is not valid for *true*DGS® probes when the DGS method is used.

4.4 Single Element Probes

After all necessary adaptation have been applied, the calculation of the transducer shape is finalized. First of all, the DGS accuracy of these new angle beam probes is of interest. The measurements were taken with a *true*DGS® probe MWB 60-4tD on an appropriate test block with flat-bottomed holes with the diameter of 3.1 mm. For each flat-bottomed hole, a back wall echo at the same sound path was measured as the reference echo. Any influence of sound attenuation is avoided by this method. Figure 4.9 demonstrates the high accuracy of the DGS evaluation of the measurements taken with a MWB 60-4tD.

The *true*DGS® transducer is curved in all directions. The resulting transducer is shown in the upper left corner of Fig. 4.10. To better visualize the curvature in the longitudinal direction, a coordinate transformation has been performed. The lower end of the transducer was shifted to the origin of the coordinate system. Then, the transducer was rotated in a way that the longitudinal axis of the transducer is fully in the x-y plane. The result is the transducer shape shown in the lower left corner of Fig. 4.10.

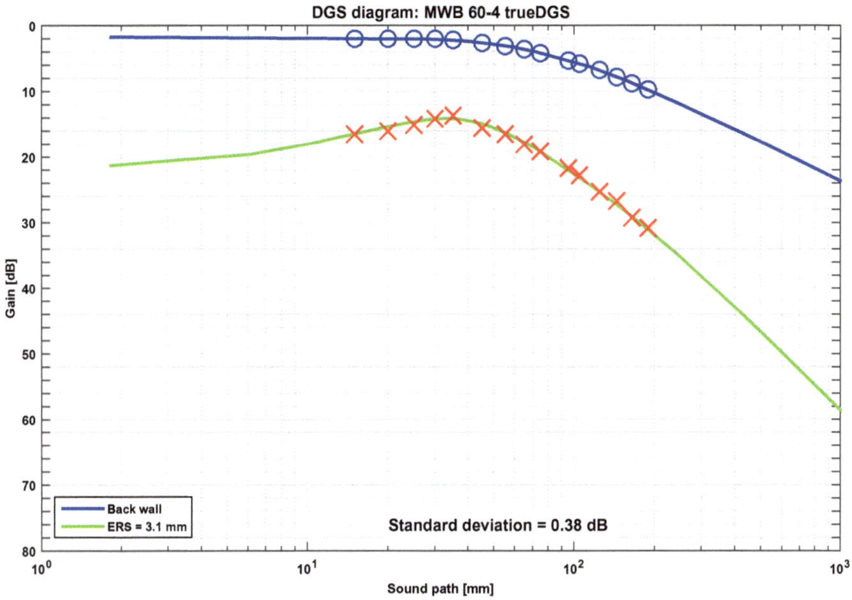

Fig. 4.9 Special DGS diagram for the MWB 60-4tD

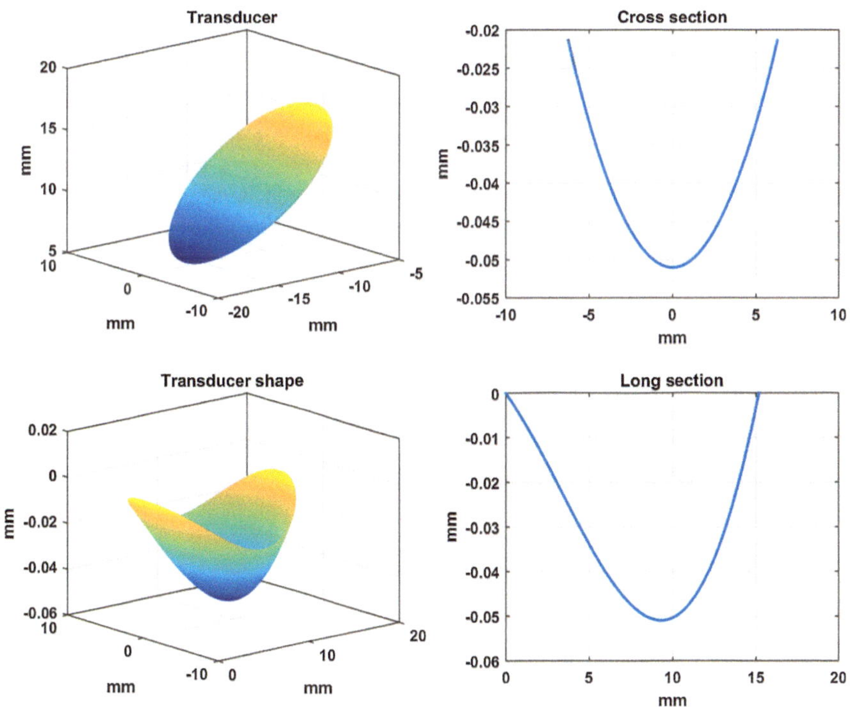

Fig. 4.10 Calculation results for the MWB 60-4tD

The transducer is curved only by a few tenth of a millimeter. But it has to be understood that the wave length is the unit of measure to be considered.

4.5 Rotational Symmetry

To investigate the rotational symmetry of this new probe technology, a CIVA simulation was carried out. The starting point was a two dimensional rectangular transducer. The elements not on the area of the calculated transducer were deactivated. The curvature was simulated using delay times. This simulation has to be understood as an approximation and not as a fully exact simulation. Figure 4.11 shows the result of this simulation. Cross sections through the sound field perpendicular to the acoustic axis are shown in distances of 0.7, 1, and 2 near field length.

The comparison with the CIVA simulation of a traditional angle beam probe with a rectangular transducer (Fig. 3.3) shows the clear improvement in the rotational symmetry.

4.5.1 Measurement of the Sound Fields

The sound fields have been measured by Salzgitter Mannesmann Forschung GmbH using the photo elastic effect [4]. The measurements were carried out on glass ($c_{shear} = 3.54$ km/s). The result of the measurement can be seen in Fig. 4.12. The symmetry to the acoustic axis is well visible in spite of the different sound velocity. Major side lobes are not recognizable.

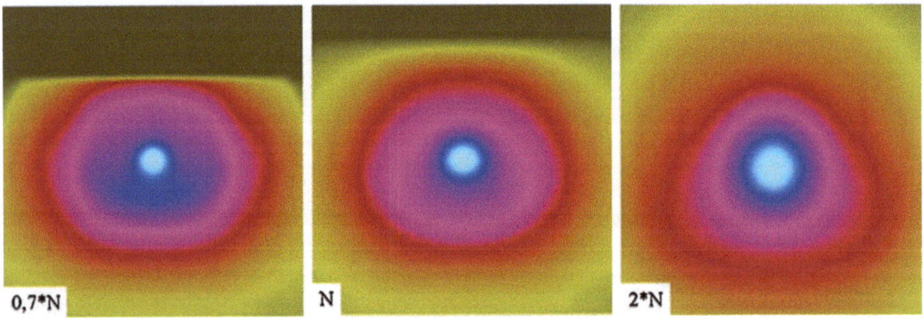

Fig. 4.11 CIVA simulation of the sound field of a *true*® probe

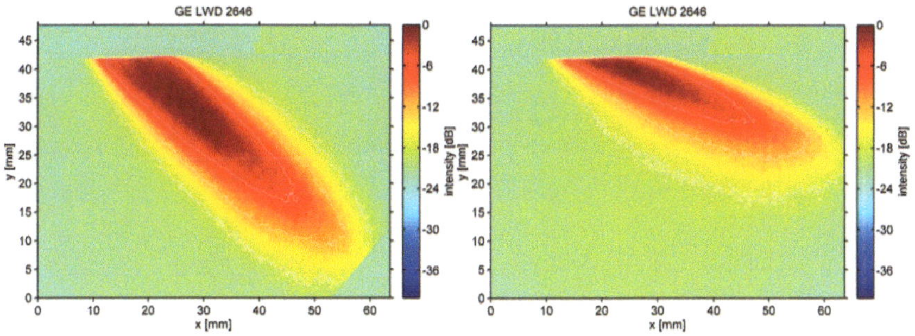

Fig. 4.12 Sound field measurements using the photo elastic effect

4.6 Advantage of the New Probe Technology

Besides the fact that *true*DGS® angle beam probes generate rotationally symmetric sound fields and have a significantly better DGS accuracy, they have another advantage

Mathematically, these probes can be handled like a straight beam probe with a circular transducer. All effects occurring at the interface between the wedge and the test piece do not have to be considered with the only exception that the zero point has to be shifted dependent on the length of the delay. This fact enables further developments like *true*DGS® phased array angle beam probes and the development of bandwidth-dependent DGS and DAC curves.

References

1. Krautkramer, J., Krautkramer, H.: Ultrasonic testing of materials. 4th Fully Revised Edition Translation of the 5th Revised German Edition. Springer, Heidelberg GmbH (1990)
2. Lavrentyev, A.I., Rokhlin, S.I.: Phase correction for ultrasonic bulk wave measurement of elastic constants in anisotropic materials. In: Thompson, D.O., Chimenti, D.E. (eds.) Review of Progress in Quantitative Nondestructive Evaluation. Plenum Press, New York (1997)
3. Wüstenberg H.: Untersuchungen zum Schallfeld von Winkelprüfköpfen für die Materialprüfung mit Ultraschall, Dissertation, BAM-Berichte Nr. 27, Berlin, August 1974
4. Schmitte T., Orth T., Spies M., Kersting T.: Schallfelder von Phased-Array Prüfköpfen: Vergleich von photoelastischen Messungen und Simulationen, DACH-Jahrestagung 2012, Graz
5. Kleinert W., Oberdörfer Y.: Präzise AVG-Bewertung mit Gruppenstrahler-Winkelprüfköpfen für alle Winkel DGZfP-Jahrestagung 2014, Potsdam. http://www.ndt.net/article/dgzfp2014/papers/p44.pdf
6. Kleinert W., Oberdörfer Y., Splitt G.: The Ideal Angle Beam Probe for DGS Evaluation, 10th ECNDT, Moskau (2010). http://www.ndt.net/article/ecndt2010/reports/1_03_64.pdf

Chapter 5
New Probe Technology, Phased Array Probes

Abstract The new probe technology can be extended to phased array angle beam probes. This new technology allows to calculate all parameters for each steering angle of the phased array probe. Not only the delay laws are determined but for each angle all other data needed are determined in this chapter. With this data a special DGS diagram can be derived from the general one for each angle making it easy to use the DGS method for sizing with phased array angle beam probes.

Long sections of the near fields for the nominal and the virtual transducer are calculated using the new probe technology. Calculations of long sections are performed with the acoustic axis going through the origin of the coordinate system. Additionally, the rim beams to the end of the near field are determined, Fig. 5.1. Further processing is different for positive and negative steering angles. Using positive steering angles, the virtual transducer is shifted mathematically in such a way that the lower edges of both transducers coincide and that the upper rim beam of the virtual transducer crosses the upper edge of the original transducer. Negative steering angles are dealt with accordingly.

In both cases, a nonlinear system of equations with the following variables has to be solved:

- diameter of the straight beam probe used to calculate the virtual transducer
- delay length of the virtual transducer
- sound exit point of the virtual transducer

This system of nonlinear equations can be solved numerically, the result is shown in Fig. 5.2. The parameters derived for the long section of the virtual transducer are used to calculate the entire transducer shape. This is necessary to derive the area of the virtual transducer. The area of the resulting transducer is required to calculate the diameter D_{dgs} of a circle with the same area, refer to Sect. 4.3.3.

Now, the following parameters are known for the virtual transducer:

- diameter of the straight beam probe used to calculate the virtual transducer (result of the system of equations)
- diameter D_{dgs} of a circle having the same area as the virtual transducer
- delay of the virtual transducer (result of the system of equations)

W. Kleinert, *Defect Sizing Using Non-destructive Ultrasonic Testing*,
DOI 10.1007/978-3-319-32836-2_5

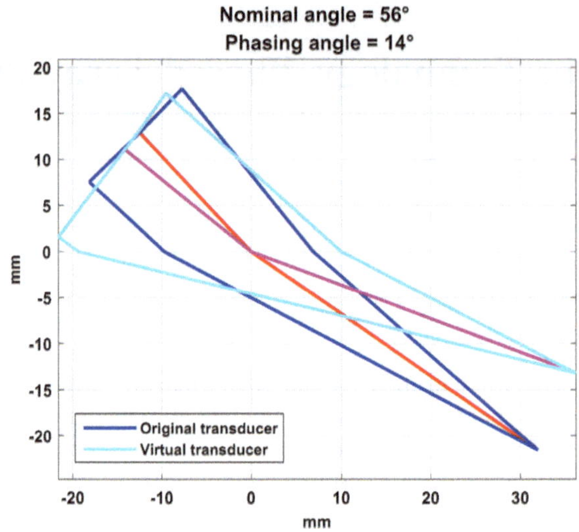

Fig. 5.1 Long sections: nominal and virtual transducer

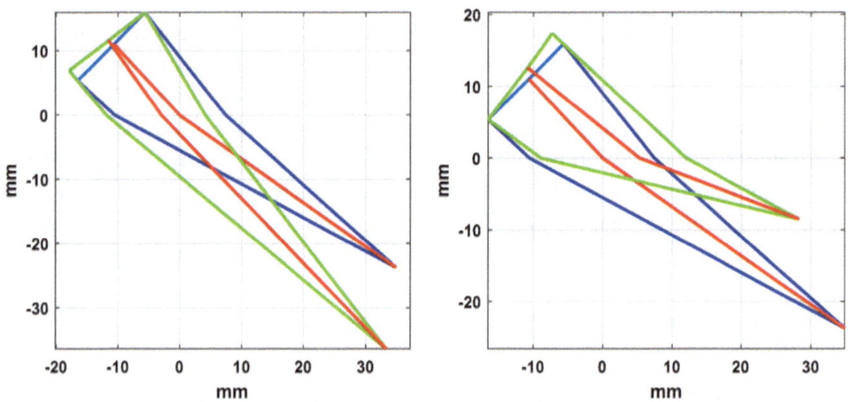

Fig. 5.2 Long sections after solving the system of equations

- sound exit point of the virtual transducer (result of the system of equations). For the sound exit points additional adaptations are required.

With this data, the DGS method can be applied to phased array angle beam probes. Only the delay laws are still missing [2].

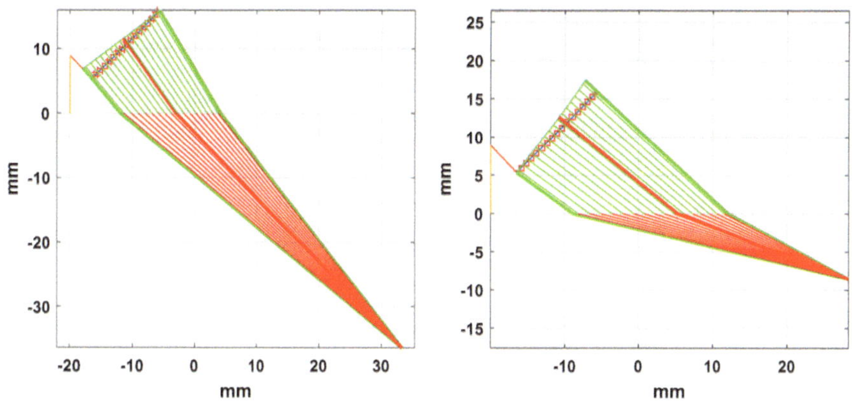

Fig. 5.3 Calculation of the delay times

5.1 Delay Laws

The center points of the elements of the original transducer are derived and connected to the end of the near field considering Snell's Law. The lines in the wedge of the probe are prolonged to the intersections with the virtual probe. The resulting distances between these intersections and the center of the elements are converted into delay times applying the sound velocity c_V in the wedge [2], Fig. 5.3.

5.2 DGS Accuracy

During the development of these new phased array angle beam probes, ten prototypes were built for 2 and 4 MHz. Test blocks with flat-bottomed holes were available for angles of incidence of 45, 53, 60, 65, and 70°. Different operators took the huge amount of measurements manually. The DGS evaluation was performed according to EN ISO 16811:2012 [1] considering only sound paths above 0.7 near field lengths. For each flat-bottomed hole, a reference echo was taken from a flat back wall at the same sound path to avoid any influences of sound attenuation. Exemplary one of the DGS diagrams including the evaluation of the measurements for 70° is shown in Fig. 5.4.

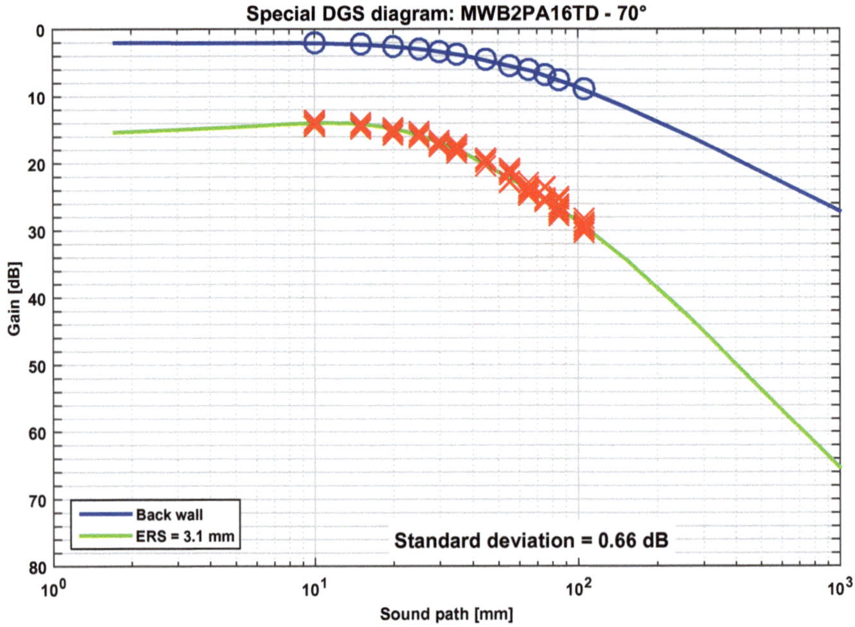

Fig. 5.4 DGS evaluation of measurements taken with a *true*DGS® phased array angle beam probe at 70°

5.3 Sound Exit Points

As documented in Table 5.1, the DGS accuracy of the *true*DGS® is significantly higher than the results of conventional probes. This is valid for single element and for phased array angle beam probes. Being able to calculate all data is an additional advantage of this new probe technology. Unfortunately, there seems to be one exception. The calculated sound exit points do not match well the measured values.

For a better understanding, the article *Gruppenlaufzeit and Bündelversetzung bei der Schrägreflexion—Auswirkungen auf die praktische Werkstoffprüfung mit Ultraschall* by Ludwig Niklas [3] is taken into account.

Table 5.1 DGS accuracy of a 2 MHz *true*DGS® phased array angle beam probe

Angle of incidence	Standard deviation (db)
45°	1.21
53°	0.73
60°	0.98
65°	0.82
70°	0.66

L. Niklas introduces for the calculation of the beam displacement the following two abbreviations:

$$s = \sin\,\alpha \text{ and } q = \frac{c_t}{c_l} \tag{5.1}$$

with:

- α: angle of incidence
- c_t: sound velocity of the transverse wave in the test material
- c_l: sound velocity of the longitudinal wave in the test material
- λ_t: wave length of the transverse wave in the test material

Using these abbreviations, the following formula is given for the beam displacement Δ by L. Niklas [3]:

$$\Delta = -\frac{\lambda_t}{\pi}\,\frac{s\left(1-2s^2\right)\left[8\left(q^2-1\right)s^4+4\left(3-q^2\right)s^2-8q^2\right]}{\sqrt{1-s^2}\sqrt{s^2-q^2}\left[\left(1-2s^2\right)^4+16s^4\left(1-s^2\right)\left(s^2-q^2\right)\right]} \tag{5.2}$$

The beam displacement can reach several wave lengths dependent on the angle of incidence. At 45°, the beam displacement is zero. Therefore, the beam displacement cannot be the only cause for the deviations between measurement and the first approach to calculate the sound exit points.

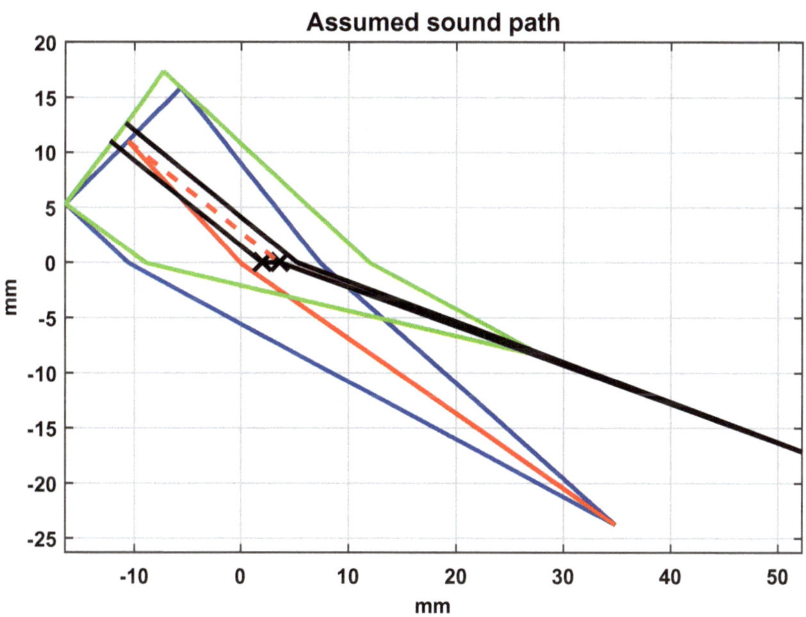

Fig. 5.5 Assumption to calculate the sound exit points

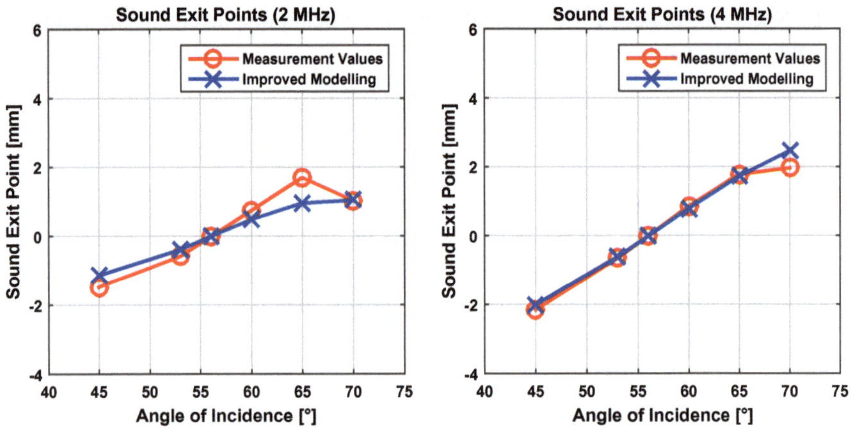

Fig. 5.6 Measured sound exit points compared to the calculated sound exit points

To explain the measured values for the sound exit points an assumption is made. Figure 5.5 shows the assumption for a sound path of 50 mm. According to the assumption, the sound follows the modeled acoustic axis to the reflector and takes the fastest path to the central point of the original transducer on the way back (dashed line in Fig. 5.5). In addition, the beam displacement according to L. Niklas takes place. Figure 5.6 shows the good match between measured sound exit points and those calculated based on the described assumption. Further validation of this assumption has to be carried out when more different phased array angle beam probes based on the new technology are available.

References

1. Ultrasonic testing - Sensitivity and range setting (ISO 16811:2012), German version EN ISO 16811:2014
2. Kleinert W., Oberdörfer Y.: Präzise AVG-Bewertung mit Gruppenstrahler-Winkelprüfköpfen für alle Winkel DGZfP-Jahrestagung 2014, Potsdam. http://www.ndt.net/article/ecndt2010/reports/1_03_64.pdf
3. Niklas L.: *Gruppenlaufzeit und Bündelversetzung bei der Schrägreflexion, Auswirkungen auf die praktische Werkstoffprüfung mit Ultraschall*, DEUTSCHER VERBAND FÜR MATERIALPRÜFUNG (DVM), Materialprüf. Bd. 7 Nr. 8, Seiten 281 bis 320, Düsseldorf (1965)

Chapter 6
New Probe Technology, Curved Coupling Surfaces

Abstract Focusing or defocusing effects occur when ultrasonic probes are coupled to curved surfaces. The new probe technology ensures clearly defined rotationally symmetric sound fields if the coupling surface can be described mathematically and the new probe technology is applied to derive the shape of the transducer. This is not only valid for single angle beam probes but as well for phased array angle beam probes. Even for very complex coupling surfaces such a probe can be designed. An example is given using a railway solid axle. The restriction in the EN ISO 16811:2012 for the use of the DGS method having curved coupling surfaces can be neglected when probes are applied according to the introduced technology.

If the coupling surface is concave, the EN ISO 16811:2012 [1] requires the matching of the probe wedge to the coupling surface unless the diameter of the test piece is large enough to ensure good coupling. For convex coupling surfaces, matching is required if the following conditions are valid depending on the nature of the curvature, Fig. 6.1[1]:

$$\begin{aligned} D_{obj} &< 10\ l_{ps} \\ D_{obj} &< 10\ w_{ps} \end{aligned} \tag{6.1}$$

with:

- D_{obj}: diameter of the test object
- l_{ps}: length of the probe shoe
- w_{ps}: width of the probe shoe

Whenever a matching of the probe shoe is required, the EN ISO 16811:2012 [1] does not allow using the DGS method for evaluation. Using the *true*DGS® technology even for curved coupling surfaces a transducer generating a rotationally symmetric sound field can be calculated as long as the coupling surface can be described mathematically. The restriction for the DGS use in the EN ISO 16811:2012 is not necessary when *true*DGS® angle beam probes are used [4].

Let the coupling surface be described by:

[1]Reproduction with permission of DIN Deutsches Institut für Normung e. V.

W. Kleinert, *Defect Sizing Using Non-destructive Ultrasonic Testing*,
DOI 10.1007/978-3-319-32836-2_6

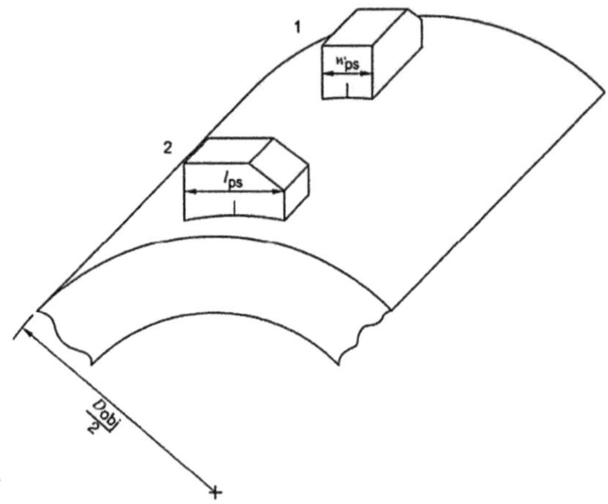

Key
1 Transversely curved
2 Longitudinally curved

Figure 1 — Length, l_{ps}, and width, w_{ps}, of probe shoe in direction of curvature of the test object

Fig. 6.1 Figure taken from EN ISO 16811:2012

$$\overline{z} = S(\overline{x}, \overline{y}) \tag{6.2}$$

where $\overline{x}, \overline{y}, \overline{z}$ are the coordinates of the point at which the sound beam under construction travels from the end of the near field into the matched wedge of the probe. With this, Eqs. (4.2) and (4.4) have to be adapted to

$$\mathbf{d} = \begin{pmatrix} \overline{x} \\ \overline{y} - y_T \\ S(\overline{x}, \overline{y}) - z_T \end{pmatrix} \tag{6.3}$$

and

$$\mathbf{v} = \begin{pmatrix} x_w - \overline{x} \\ y_w - \overline{y} \\ S(\overline{x}, z_w - \overline{y}) \end{pmatrix} \tag{6.4}$$

6.1 Fastest Path

Here, the shortest time from a point $W(x_w, y_w, z_w)$ on the transducer to the transition point $D(\overline{x}, \overline{y}, \overline{z})$ to the end of the near field $T(0, y_T, z_T)$ is derived. To find the shortest time the partial derivatives with respect to $\overline{x}$ and $\overline{y}$ are calculated

$$\frac{\partial t}{\partial \overline{x}} = -\frac{1}{c_v} \frac{x_w - \overline{x} + (z_w - S(\overline{x}, \overline{y})) \frac{\partial S}{\partial \overline{x}}}{\sqrt{(x_w - \overline{x})^2 + (y_w - \overline{y})^2 + (z_w - S(\overline{x}, \overline{y}))^2}} + \frac{1}{c_m} \frac{\overline{x} + (S(\overline{x}, \overline{y}) - z_T)^2 \frac{\partial S}{\partial \overline{x}}}{\sqrt{\overline{x}^2 + (\overline{y} - y_T)^2 + (S(\overline{x}, \overline{y}) - z_T)^2}} \tag{6.5}$$

$$\frac{\partial t}{\partial \overline{y}} = -\frac{1}{c_v} \frac{y_w - \overline{y} + (z_w - S(\overline{x}, \overline{y})) \frac{\partial S}{\partial \overline{y}}}{\sqrt{(x_w - \overline{x})^2 + (y_w - \overline{y})^2 + (z_w - S(\overline{x}, \overline{y}))^2}} + \frac{1}{c_m} \frac{\overline{y} - y_T + (S(\overline{x}, \overline{y}) - z_T) \frac{\partial S}{\partial \overline{y}}}{\sqrt{\overline{x}^2 + (\overline{y} - y_T)^2 + (S(\overline{x}, \overline{y}) - z_T)^2}} \tag{6.6}$$

For further calculations, the following two abbreviations for the partial derivatives of the coupling surface $S(\overline{x}, \overline{y})$ with respect to $\overline{x}$ and $\overline{y}$ are used:

$$m_x = \frac{\partial S}{\partial \overline{x}} \quad \text{and} \quad m_y = \frac{\partial S}{\partial \overline{y}} \tag{6.7}$$

The vectors $\mathbf{e_{tx}}$ and $\mathbf{e_{ty}}$ spanning the tangent plane are introduced

$$\mathbf{e_{tx}} = \begin{pmatrix} 1 \\ 0 \\ m_x \end{pmatrix} \tag{6.8}$$

$$\mathbf{e_{ty}} = \begin{pmatrix} 0 \\ 1 \\ m_y \end{pmatrix} \tag{6.9}$$

Hence, Eqs. (6.5) and (6.6) can be simplified

$$\frac{\partial t}{\partial \overline{x}} = -\frac{|\mathbf{e_{tx}}|}{c_v} \frac{\mathbf{e_{tx}} \cdot \mathbf{v}}{|\mathbf{e_{tx}}|\,|\mathbf{v}|} + \frac{|\mathbf{e_{tx}}|}{c_m} \frac{\mathbf{e_{tx}} \cdot \mathbf{d}}{|\mathbf{e_{tx}}|\,|\mathbf{d}|} \tag{6.10}$$

$$\frac{\partial t}{\partial \overline{y}} = -\frac{|\mathbf{e_{ty}}|}{c_v} \frac{\mathbf{e_{ty}} \cdot \mathbf{v}}{|\mathbf{e_{ty}}|\,|\mathbf{v}|} + \frac{|\mathbf{e_{ty}}|}{c_m} \frac{\mathbf{e_{ty}} \cdot \mathbf{d}}{|\mathbf{e_{ty}}|\,|\mathbf{d}|} \tag{6.11}$$

To derive the minimum of time t, the two equations are set to zero

$$\frac{1}{c_v} \frac{\mathbf{e_{tx}} \cdot \mathbf{v}}{|\mathbf{e_{tx}}|\,|\mathbf{v}|} = \frac{1}{c_m} \frac{\mathbf{e_{tx}} \cdot \mathbf{d}}{|\mathbf{e_{tx}}|\,|\mathbf{d}|} \tag{6.12}$$

$$\frac{1}{c_v} \frac{\mathbf{e_{ty}} \cdot \mathbf{v}}{|\mathbf{e_{ty}}|\,|\mathbf{v}|} = \frac{1}{c_m} \frac{\mathbf{e_{ty}} \cdot \mathbf{d}}{|\mathbf{e_{ty}}|\,|\mathbf{d}|} \tag{6.13}$$

6.2 Angles

The following variables are introduced:

- α_{tx}: angle between sound beam and the tangent plane in x-direction in the probe wedge
- α_{ty}: angle between sound beam and the tangent plane in y-direction in the probe wedge
- β_{tx}: angle between sound beam and the tangent plane in x-direction in the test object
- β_{ty}: angle between sound beam and the tangent plane in y-direction in the test object

With these variables, Eqs. (6.12) and (6.13) can be written as

$$\frac{\cos(\alpha_{tx})}{\cos(\beta_{tx})} = \frac{c_v}{c_m} \tag{6.14}$$

and

$$\frac{\cos(\alpha_{ty})}{\cos(\beta_{ty})} = \frac{c_v}{c_m} \tag{6.15}$$

In the next step, the angles to the z-axis are derived. Therefore, the vector $\mathbf{e_{tz}}$ perpendicular to the tangent plane is derived. The vector $\mathbf{e_{tz}}$ is the result of the vector product of $\mathbf{e_{ty}}$ and $\mathbf{e_{tx}}$

$$\mathbf{e_{tz}} = \begin{pmatrix} 0 \\ 1 \\ m_y \end{pmatrix} \times \begin{pmatrix} 1 \\ 0 \\ m_x \end{pmatrix} = \begin{pmatrix} m_x \\ m_y \\ -1 \end{pmatrix} \tag{6.16}$$

Hence, for the angle β_{tz} between the sound beam and the z-axis in the test object follows:

$$\cos(\beta_{tz}) = \frac{\mathbf{e_{tz}} \cdot \mathbf{d}}{|\mathbf{e_{tz}}|\ |d|} \tag{6.17}$$

The angles α_{tx} and α_{ty} are defined to the x- and y-axes and are not as usual to the perpendicular lines. From this results, the ratio of the cosine functions are in Eqs. (6.14) and (6.15). In contrast, the angle β_{tz} is given to the z-axis. Hence, for the angle α_{tz} in the probe wedge follows:

$$\frac{\sin(\alpha_{tz})}{\sin(\beta_{tz})} = \frac{c_v}{c_m} \tag{6.18}$$

The angles β_{tx} and β_{ty} result from

$$\cos(\beta_{tx}) = \frac{\mathbf{e_{tx}} \cdot \mathbf{d}}{|\mathbf{e_{tx}}|\ |\mathbf{d}|} \tag{6.19}$$

and

$$\cos\left(\beta_{ty}\right) = \frac{\mathbf{e_{ty}} \cdot \mathbf{d}}{\left|\mathbf{e_{ty}}\right| \; |\mathbf{d}|} \tag{6.20}$$

Hence, all angles β_{tx}, β_{ty}, and β_{tz} are known.

6.3 Transducer Coordinates

Let vector $\mathbf{v}^T = (v_1, v_2, v_3)$ with $|\mathbf{v}| = 1$ be the vector in the probe wedge defined by the angles α_{tx}, α_{ty}, and α_{tz}. These angles can be determined using the following dot products:

$$\begin{aligned} \mathbf{e_{tx}} \cdot \mathbf{v} &= |\mathbf{e_{tx}}| \; \cos(\alpha_{tx}) \\ \mathbf{e_{ty}} \cdot \mathbf{v} &= \left|\mathbf{e_{ty}}\right| \; \cos\left(\alpha_{ty}\right) \\ \mathbf{e_{tz}} \cdot \mathbf{v} &= |\mathbf{e_{tz}}| \; \cos(\alpha_{tz}) \end{aligned} \tag{6.21}$$

This system of equations can be written as

$$\begin{pmatrix} 1 & 0 & m_x \\ 0 & 1 & m_y \\ m_x & m_y & -1 \end{pmatrix} \begin{pmatrix} v_1 \\ v_2 \\ v_3 \end{pmatrix} = \begin{pmatrix} \cos(\alpha_{tx}) \; \sqrt{1+m_x^2} \\ \cos\left(\alpha_{ty}\right) \; \sqrt{1+m_y^2} \\ \cos(\alpha_{tz}) \; \sqrt{1+m_x^2+m_y^2} \end{pmatrix} \tag{6.22}$$

Since the angles α_{tx}, α_{ty}, and α_{tz} are known from Eqs. (6.14), (6.15), and (6.18), the system of equations (6.22) can be solved.

Based on the time of flight of the sound beam under construction the length of the delay can be derived. First, the time of flight $t(\gamma)$ in the predefined straight beam probe is determined using Eqs. (4.12) and (4.14). The time of flight t_m in the test object follows from vector $\mathbf{d}$ and the sound velocity. Hence, the time of flight in the wedge is

$$\begin{aligned} t_m &= \frac{|\mathbf{d}|}{c_m} \\ t_v &= t(\gamma) - t_m \end{aligned} \tag{6.23}$$

The length of vector $\mathbf{v}$ equals 1. This vector is now multiplied with the calculated length of the delay

$$\mathbf{w} = c_v \; t_v \begin{pmatrix} v_1 \\ v_2 \\ v_3 \end{pmatrix} \tag{6.24}$$

The vector from the origin of the coordinate system to the puncture point on the coupling surface $D(\overline{x}, \overline{y}, \overline{z})$ is given by

$$\begin{pmatrix} \overline{x} \\ \overline{y} \\ \overline{z} \end{pmatrix}$$

The sum of this vector and vector **w** is the vector from the origin of the coordinate system to the transducer point $W(x_w, y_w, z_w)$

$$\begin{pmatrix} x_w \\ y_w \\ z_w \end{pmatrix} = \begin{pmatrix} \overline{x} \\ \overline{y} \\ \overline{z} \end{pmatrix} + c_v\, t_v \begin{pmatrix} v_1 \\ v_2 \\ v_3 \end{pmatrix} \tag{6.25}$$

With this the transducer point W is known. Repeating this calculation for all $\overline{x}$ and $\overline{y}$ complying with the condition $\gamma \leq \gamma_{\max}$ results in the entire point cloud defining the transducer shape.

6.4 Example: Solid Axle

The design of a transducer for curved surfaces is described using a solid train axle from Deutsche Bahn with the identifier BA013 as an example [2, 3]. For testing the axle ultrasonically, only the coupling area shown in Fig. 6.2 is available. There is not a single space with a constant diameter of the axle in this area.

Having such a coupling geometry, the angle to the axis of the axle is more important than the angle of incidence. Hence, the angle to the axis has to be chosen first to start the design of the transducer. For this example, the following parameters are used:

- frequency: 4 MHz
- transducer diameter (straight beam probe): 22 mm
- angle to the axis of the solid axle: 50°
- delay: 17 mm
- sound exit point: 7.5 mm

First, the z-coordinate z_0 for the selected sound exit point of $y_0 = 7.5$ mm is derived and the slope of the geometry at this point is calculated. In this example, the slope of the tangent at the sound exit point is $m_s = 0.1006$.

The near field length N for the selected diameter D is calculated. The remaining segment to the end of the near field of the angle beam probe under construction is derived based on the selected delay length. This segment is drawn from the sound exit point under consideration of the selected angle to the axis of the solid axle. The angle of incidence is given by the difference of the angle to the axis and the slope at the sound exit point. In this example, the angle of incidence results in 44.26°.

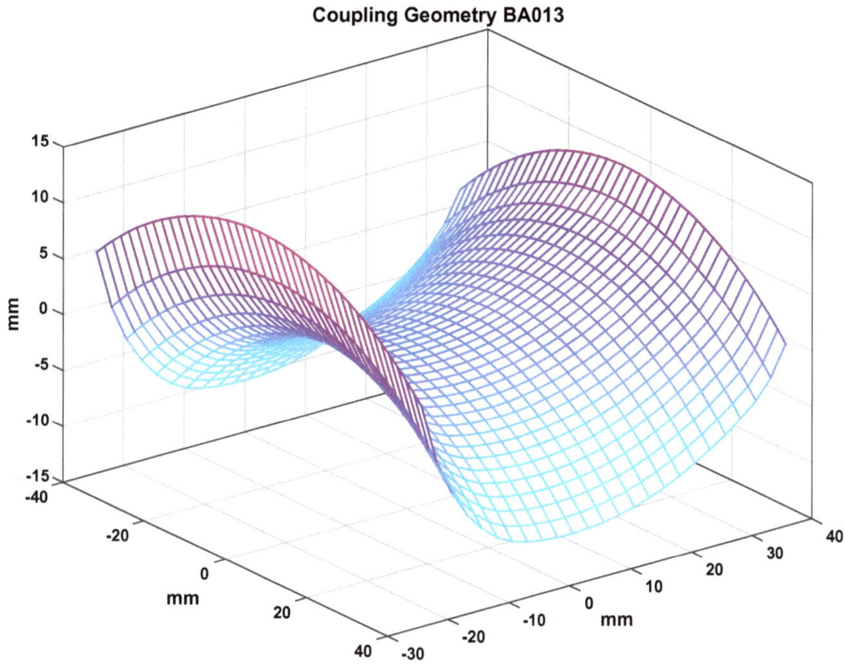

Fig. 6.2 Coupling geometry of the solid axle BA013

Hence, the delay line can be drawn into the sketch Fig. 6.3. Now, the coordinates y_c and z_c of the transducer point W and the coordinates y_T and z_T of the end of the near field T are known. Now, the complete transducer shape can be calculated. The

Fig. 6.3 Coupling geometry and acoustic axis

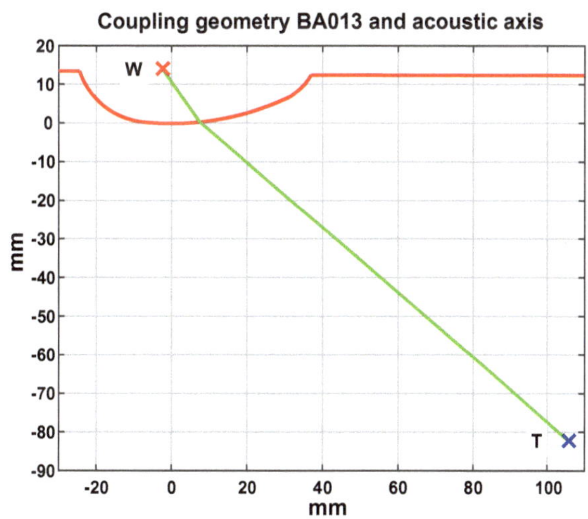

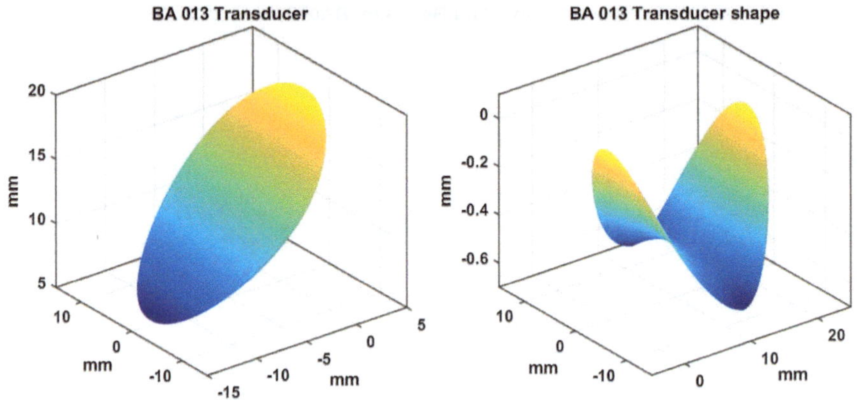

Fig. 6.4 Transducer and transducer shape

angle of incidence and the length of the delay have to be derived for each sound beam considering the coupling geometry. Figure 6.4 shows the resulting transducer and the transducer shape.

6.5 Delay Laws

The procedure described in Sect. 5.1 for calculating the delay laws does not work when the cut of the sound field with the coupling surface is not a straight line. In these cases, every shift of the sound beam results in a new angle of incidence. Hence, for these kinds of coupling geometries a new procedure for calculating the delay laws had to be developed. Here has to be distinguished between positive and negative steering angles as well. The following description is for steering angles greater than the nominal angle. The procedure for negative steering angles works accordingly.

For the calculation of the virtual transducer the following steps are repeated until a predefined accuracy has been reached. The following steps have been implemented in a while-loop:

(1) A step size is defined (in the beginning for example 10 mm).
(2) Starting point of the calculation is the sound exit point of the original transducer.
(3) For the planned angle to the axis of the solid axle from this exit point the end of the potential near field is calculated. The sound path from the sound exit point to the end of the potential near field is a multiple of the step size (depending on the number of repetitions).
(4) The fastest path from the end of the potential near field to the lower edge of the original transducer is calculated including the intersection of this rim beam with the coupling surface.

(5) The fastest path from the end of the potential near field to the upper edge of the original transducer including the intersection of this rim beam with the coupling surface is derived.
(6) The angle bisector between the two rim beams is derived. The intersection of the angle bisector with the coupling surface is the new starting point for the calculation of the potential near field in the next repetition of the loop (the new sound exit point). This results in the new angle of incidence.
(7) The angle included by the rim beams is determined.
(8) The total time of flight t_{total} from the lower edge of the original transducer to the end of the potential near field is calculated under consideration of the angle dependent phase shift.
(9) If the correct near field end would have been reached already, the near field length would be given by: $(t_{total} - T/2)\ c_m$ with $T = 1/f$ and c_m sound velocity in the test object and f the frequency of the probe used.
(10) For this derived potential near field length, the angle of the beam spread is calculated for straight beam insonification. This angle is compared to the included angle derived in step 7.
(11) If the angle of beam spread is smaller than the one calculated in step 7 the sound path in step 3 is prolonged by the step size and the complete procedure is started all over again.
(12) If the angle calculated for the potential near field length is larger for the first time than the one calculated in step 7, the sound path is reduced by one step size and the step size is divided by 10.
(13) This procedure is repeated until the predefined accuracy is reached.

The result of this procedure can be seen in Fig. 6.5. For deriving the delay laws, the center point of the elements of the original transducer are connected to the end of the near field considering Snell's Law. For each sound beam, the angle of incidence has to be calculated based on the curvature of the coupling surface. The connecting lines are prolonged to the intersection with the virtual transducer. The distances between original and virtual transducer are the base for calculating the delay laws applying the sound velocity in the delay material.

All focusing or defocusing effects from the curved coupling surface are eliminated by applying this technology. Even in these cases, rotationally symmetric sound fields are generated.

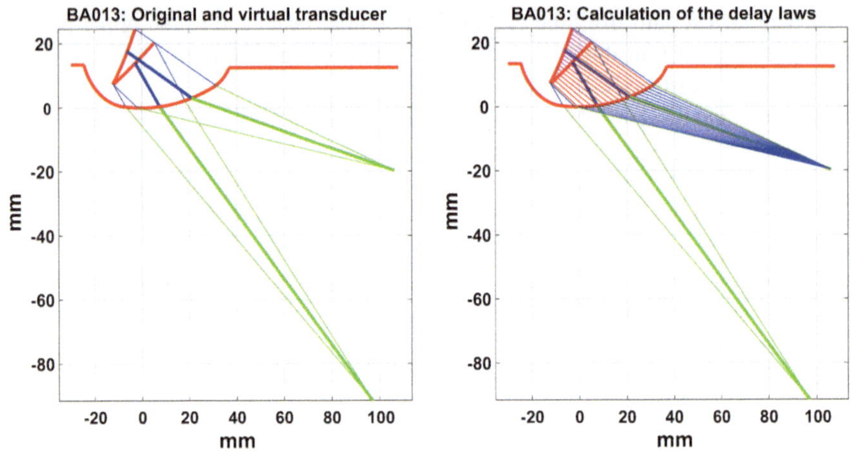

Fig. 6.5 Original and virtual transducer including calculation of the delay laws for the solid axle BA 013

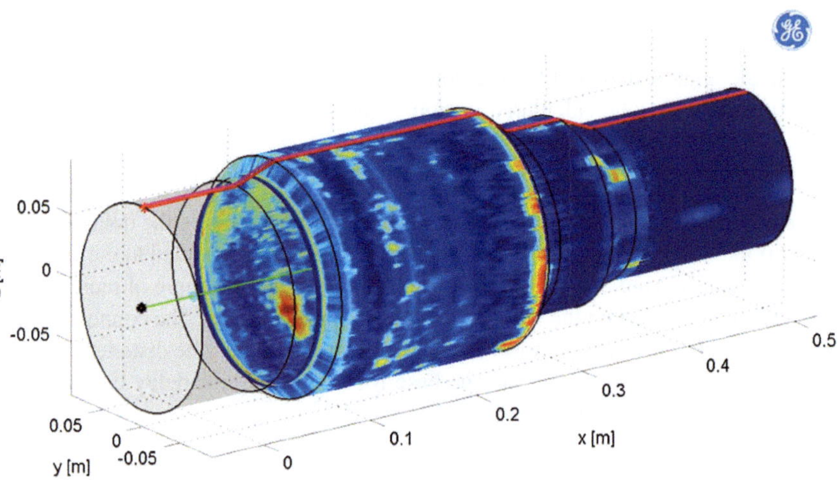

Fig. 6.6 Solid axle inspected ultrasonically

Figure 6.6 shows a solid axle tested ultrasonically using a *true*DGS® phased array angle beam probe calculated for this coupling geometry particularly [2, 3]. A program utilized to calculate delay laws for complex coupling geometries is illustrated in Fig. 6.7.

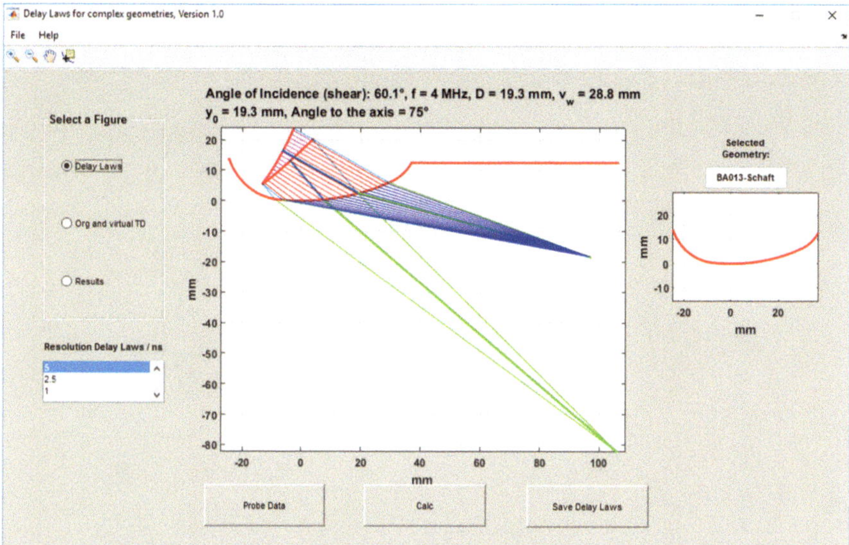

Fig. 6.7 Tool for calculating delay laws for complex geometries

References

1. Ultrasonic testing—Sensitivity and range setting (ISO 16811:2012); German version EN ISO 16811:2014
2. Kleinert W., Chinta P.: Neues Ultraschallverfahren zur Prüfung von Vollwellen 8. Fachtagung ZfP im Eisenbahnwesen, Wittenberge (2014).http://www.ndt.net/article/dgzfp-misc/rail2014/papers/13.pdf
3. Kleinert W., Chinta P.: Automatisierte Prüfung von Eisenbahnvollwellen unter besonderer Berücksichtigung der Geometrieeinflüsse. DGZP-Jahrestagung, Potsdam (2014). http://www.ndt.net/article/dgzfp2014/papers/mi3a2.pdf
4. Kleinert W., Oberdörfer Y.: Präzise AVG-Bewertung mit Gruppenstrahler-Winkelprüfköpfen für alle Winkel. DGZfP-Jahrestagung, Potsdam (2014). http://www.ndt.net/article/ecndt2010/reports/1_03_64.pdf

Chapter 7
Bandwidth-Dependent DGS Diagrams

Abstract The advantage of the new probe technology is the fact that these angle beam probes behave like straight beam probes with circular transducers for transverse waves. Therefore, it is much easier to handle these probes mathematically. Without having to consider all the complex effects at the interface between wedge and test piece, bandwidth-dependent DGS diagrams can be derived. Even a general DGS diagram can be developed bandwidth dependently for families of single element angle beam probes or families of phased array angle beam probes as long as they have all nearly the same bandwidth. With these bandwidth-dependent DGS diagrams, the restriction in the EN ISO 16811:2012 to use only sound paths larger than 0.7 near field lengths can be neglected.

The DGS accuracy of the probes based on the new technology is significantly higher than the accuracy of conventional angle beam probes. Figure 7.1 illustrates the DGS evaluation of measurements taken with a *true*DGS® phased array probe according to the EN ISO 16811:2012. In this standard, the DGS evaluation is allowed for sound paths >0.7 N only. This working range is shown in Fig. 7.1 accordingly.

In the range below 0.7 N, reflectors are oversized when evaluated using the general DGS diagram published in the EN ISO 16811:2012, Fig. 7.2.

Bandwidth-dependent DGS diagrams will enable the DGS evaluation for the entire range of sound paths.

7.1 Single Frequency Ultrasound

Before the bandwidth-dependent DGS diagram is introduced, some thoughts are given to the first development of the DGS diagram. Therefore, the sound pressure on the acoustic axis generated by a straight beam probe with a flat circular transducer is examined for single frequency ultrasound.

W. Kleinert, *Defect Sizing Using Non-destructive Ultrasonic Testing*,
DOI 10.1007/978-3-319-32836-2_7

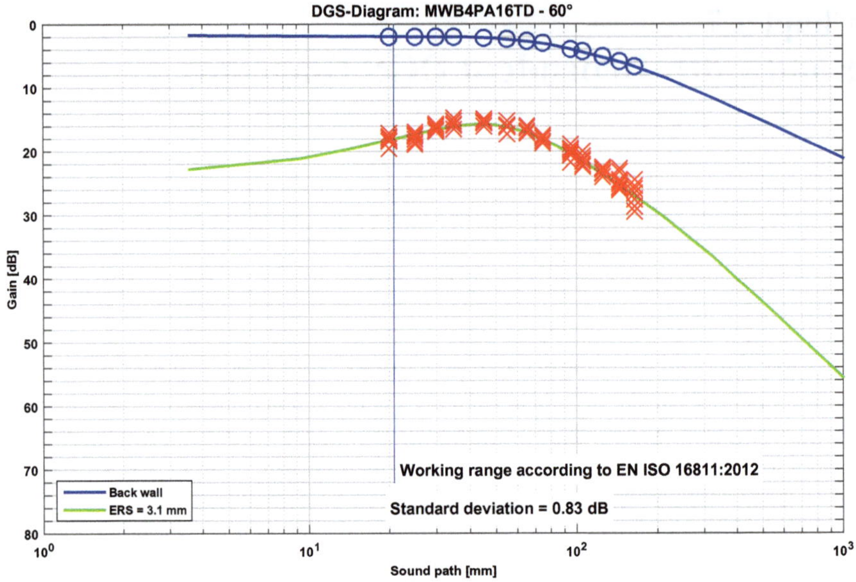

Fig. 7.1 Evaluation of measurements taken with a *true*DGS® phased array angle beam probe according to EN ISO 16811:2012

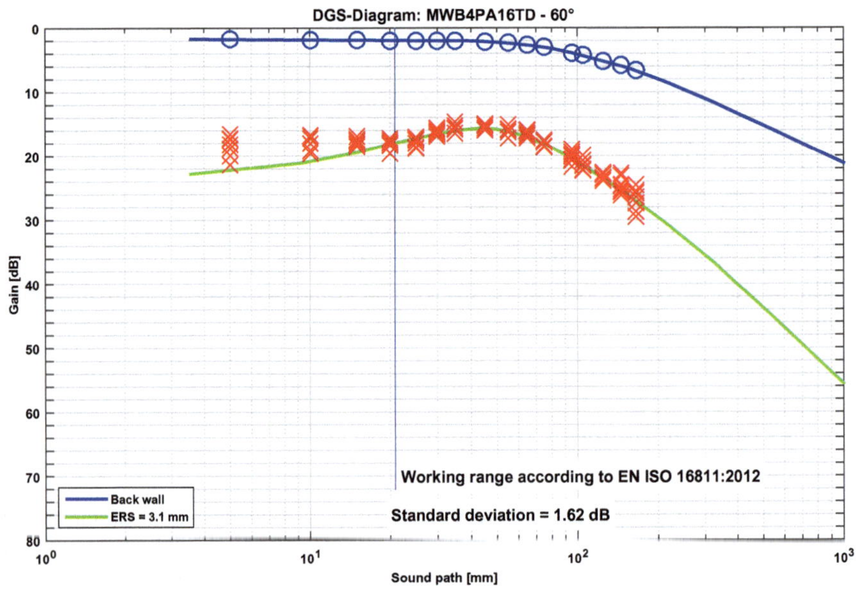

Fig. 7.2 Reflectors are oversized at sound paths below 0.7 N using the general DGS diagram from the EN ISO 16811:2012

The sound pressure p_A on the acoustic axis in a distance z at time t is given by

$$p_A(z,t) = \omega\, Z \int\limits_S \frac{1}{r}\, \cos(\omega t - kr)\, dS \tag{7.1}$$

with:

- r: distance
- ω: angular frequency
- Z: acoustic impedance
- k: angular wavenumber
- z: distance
- S: area of the circular transducer

and

$$r = \sqrt{z^2 + a^2} \;\; \text{and} \;\; k = \frac{2\,\pi}{\lambda} \tag{7.2}$$

$$dS \approx a\; da\; d\varphi \tag{7.3}$$

The relations are illustrated in Figs. 7.3 and 7.4. Hence, Eq. (7.1) can be written as

$$p_A(z,t) = 2\,\omega\, Z \int\limits_{a=0}^{\frac{D}{2}} \int\limits_{\varphi=0}^{2\pi} \frac{a}{\sqrt{a^2+z^2}}\, \cos\left(\omega t - k\,\sqrt{a^2+z^2}\right)\, da\; d\varphi \tag{7.4}$$

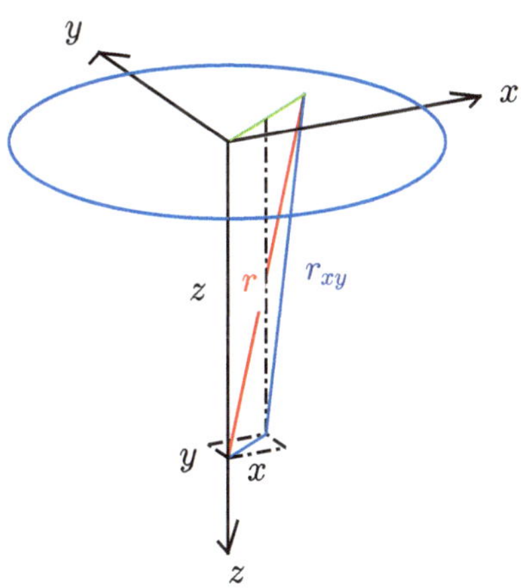

Fig. 7.3 Sketch for the calculation of the sound pressure

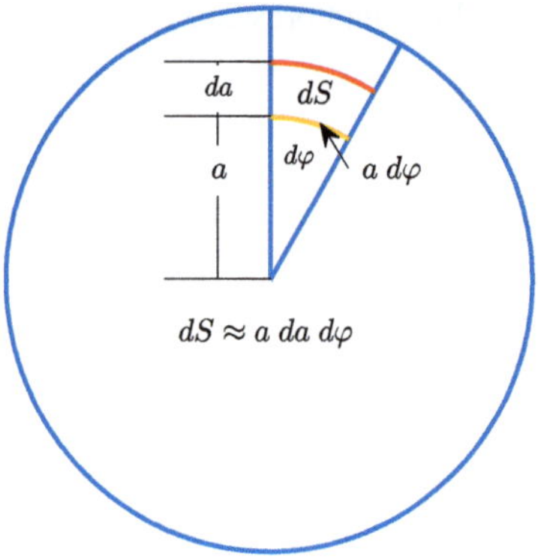

Fig. 7.4 Sketch of the circular transducer

This integral equation can be solved algebraically

$$p_A(z,t) = -2\pi c Z \left\{ \sin\left(\omega t - k\sqrt{\left(\frac{D}{2}\right)^2 + z^2}\right) - \sin\left(\omega t - kz\right) \right\} \tag{7.5}$$

Using the following trigonometric addition theorems this term can be simplified:

$$\sin\ x - \sin\ y = 2\cos\frac{x+y}{2}\ \sin\frac{x-y}{2} \tag{7.6}$$

Hence, (7.5) can be written as

$$\begin{aligned} p_A(z,t) &= 4\pi c Z \cos\frac{2\omega t - k\sqrt{\left(\frac{D}{2}\right)^2 + z^2} - kz}{2}\ \sin k\,\frac{\sqrt{\left(\frac{D}{2}\right)^2 + z^2} - z}{2} \\ &= 4\pi c Z \cos\frac{2\omega t - k\sqrt{\left(\frac{D}{2}\right)^2 + z^2} - kz}{2}\ \sin\frac{\pi}{\lambda}\left(\sqrt{\left(\frac{D}{2}\right)^2 + z^2} - z\right) \end{aligned} \tag{7.7}$$

Only the time-independent extrema of the sound pressure are of interest. Therefore, only the extrema of the cosine function are taken into account, resulting in

$$p_{\max}(z) = 4\pi c Z \left|\sin\frac{\pi}{\lambda}\left(\sqrt{\left(\frac{D}{2}\right)^2 + z^2} - z\right)\right| \tag{7.8}$$

Figure 7.5 illustrates the sound pressure on the acoustic axis calculated for a single frequency using Eq. (7.8). For a better recognizability, a semi logarithmic scale has been applied. The following parameters were used for the calculation:

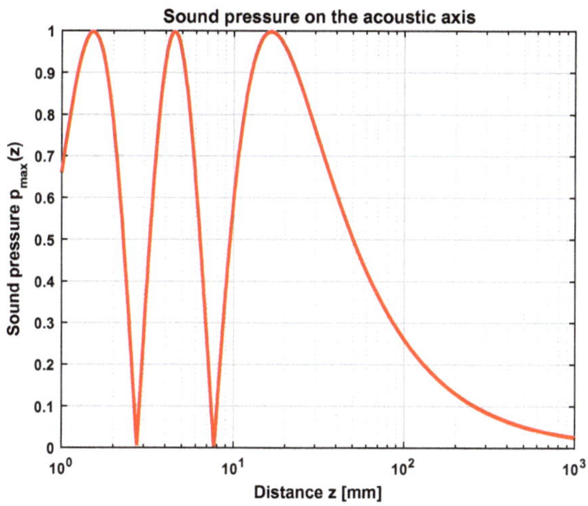

Fig. 7.5 Sound pressure on the acoustic axis calculated for a single frequency

- sound velocity $c = 5{,}920\,\mathrm{m/s}$
- frequency: $4\,\mathrm{MHz}$
- transducer diameter: $D = 10\,\mathrm{mm}$.

7.1.1 Near Field Length

The distance from the transducer to the last maximum of the sound pressure on the acoustic axis is called near field length N. The extrema can be derived using the argument of the sine function of Eq. (7.8). Generally, the sine function has extrema at

$$\frac{2n+1}{2}\pi \quad \text{with} \quad n \in \mathbb{N}_0 \tag{7.9}$$

Hence, the following term has to be evaluated:

$$\frac{\pi}{\lambda}\left(\sqrt{\left(\frac{D}{2}\right)^2 + z^2} - z\right) = \frac{2n+1}{2}\pi \quad \text{mit} \quad n \in \mathbb{N}_0 \tag{7.10}$$

$$\Rightarrow 2\sqrt{\left(\frac{D}{2}\right)^2 + z^2} = (2n+1)\lambda + 2z$$

By squaring the equation and dissolving the variable z results in

$$z = \frac{D^2 - (2n+1)^2\,\lambda^2}{4\,(2n+1)\,\lambda} \tag{7.11}$$

Hence, z has a maximum when both the expression behind the minus sign and the denominator are as small as possible. This is fulfilled for $n = 0$. With this and $z_{max} = N$ follows:

$$N = \frac{D^2 - \lambda^2}{4\,\lambda} \tag{7.12}$$

This equation is often simplified, based on the fact that generally $D \gg \lambda$ is valid, to:

$$N \approx \frac{D^2}{4\,\lambda} \tag{7.13}$$

The formula for the calculation of the near field length is valid for single frequency only. This formula is still in use today and is applied to ultrasonic pulses with a certain bandwidth. In Sect. 7.2, it will be derived that the near field length is additionally dependent on the bandwidth.

From Fig. 7.6 can be read (Pythagoras)

$$r^2 = N^2 + \left(\frac{D}{2}\right)^2 \tag{7.14}$$

By replacing N using Eq. (7.12) follows:

$$r = N + \frac{\lambda}{2} \tag{7.15}$$

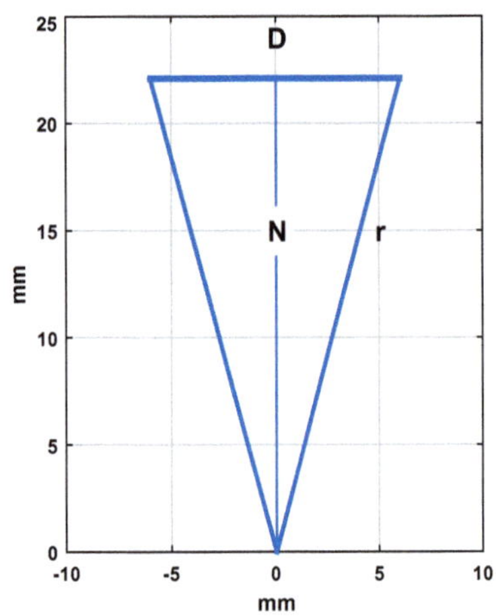

Fig. 7.6 Transducer and rim beams to the end of the near field

The fact that the sound path difference between the sound traveling along the acoustic axis to the end of the near field and the sound traveling along the rim beam amounts to $\lambda/2$ is derived for single frequency ultrasound as well.

7.2 Multi-frequency Ultrasound

Multi-frequency ultrasound means using short pulses. The length of the pulse is given by its spectrum in the frequency domain, Fig. 7.7. For the calculation, a Gaussian for the given relative bandwidth of 30 % is derived in the frequency domain. By transferring to the time domain the pulse is retrieved. In the time domain another Gaussian is determined as envelope for the pulse. This Gaussian will model the pulses in all further calculations.

7.2.1 *Near Field Length*

The near field length of this pulse, defined as last maximum of the sound pressure on the acoustic axis, is calculated numerically using the following values:

- sound velocity: $c = 5{,}920\,\text{m/s}$
- transducer diameter: $D = 20\,\text{mm}$
- frequency: $F = 4\,\text{MHz}$
- relative bandwidth: 30 %

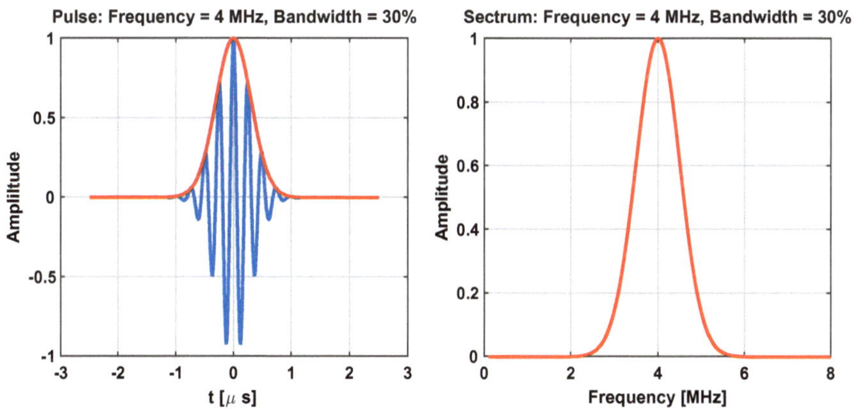

Fig. 7.7 Pulse and spectrum of multi frequency ultrasound

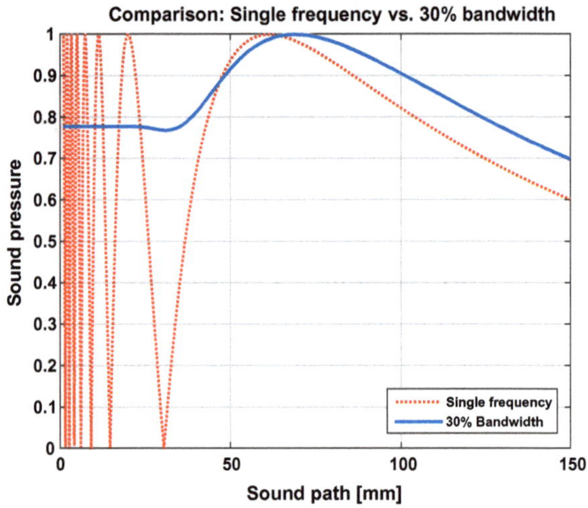

Fig. 7.8 Sound pressure calculation for a single frequency and for a pulse with a relative bandwidth of 30 %

The result of the calculation is illustrated in Fig. 7.8. The maximum and with this the near field end is at 85.4 mm. If the near field length is calculated using formula (7.12), the result is 67.2 mm. The simplified Eq. (7.13) results in 67.6 mm.

Figure 7.8 illustrates the difference between a back wall curve calculated for a single frequency according to Eq. (7.8) and a back wall curve calculated for a pulse with a relative bandwidth of 30 %.

Hence: The near field length is dependent on the bandwidth of the pulse used [1] (Table 7.1).

The integral (7.1) is modified for the calculation of the sound pressure on the acoustic axis

$$p_A(z,t) = 4\pi\omega Z \int_{a=0}^{\frac{D}{2}} e^{A\left(z-\sqrt{z^2+a^2}\right)^2} \frac{a}{\sqrt{a^2+z^2}} \cos\left(\omega t - k\sqrt{a^2+z^2}\right) da \tag{7.16}$$

Table 7.1 Near field length in dependence on the bandwidth

Bandwidth (%)	Near field length (mm)
20	78.1
30	85.4
40	91.7
50	97.2

The Gaussian travels with the sound, modeling the pulse. The Eq. (7.16) can be solved numerically only. The factor A defines the bandwidth of the pulse. Again, the trigonometric addition theorems are used. Hence, Eq. (7.16) can be split into two integrals

$$p_A(z,t) = 4\pi\omega Z\cos(\omega t)\underbrace{\int\limits_{a=0}^{\frac{D}{2}} e^{A\left(z-\sqrt{z^2+a^2}\right)^2}\frac{a}{\sqrt{a^2+z^2}}\cos\left(k\sqrt{a^2+z^2}\right)\,da}_{x_1}$$
$$+\,4\pi\omega Z\sin(\omega t)\underbrace{\int\limits_{a=0}^{\frac{D}{2}} e^{A\left(z-\sqrt{z^2+a^2}\right)^2}\frac{a}{\sqrt{a^2+z^2}}\sin\left(k\sqrt{a^2+z^2}\right)\,da}_{x_2} \tag{7.17}$$

The result for $p_A(z,t)$ is a term of the following form (without the proportionality factor $4\pi\omega Z$):

$$x_1\cos(\omega t) + x_2\sin(\omega t) \tag{7.18}$$

This term can be transferred to

$$x_1\cos(\omega t) + x_2\sin(\omega t) = B\sin(\omega t + \varphi) \tag{7.19}$$

with the unknown variables B and φ. To calculate B and φ the following trigonometric addition theorem is applied:

$$\sin(\omega\, t + \varphi) = \sin(\omega\, t)\;\cos(\varphi) + \cos(\omega\, t)\;\sin(\varphi) \tag{7.20}$$

Using this addition theorem and Eq. (7.19) results in

$$x_1\cos(\omega t) + x_2\sin(\omega t) = B\sin(\omega t)\;\cos(\varphi) + B\cos(\omega t)\;\sin(\varphi) \tag{7.21}$$

By comparison follows:

$$\begin{aligned} x_1 &= B\sin(\varphi) \\ x_2 &= B\cos(\varphi) \end{aligned} \tag{7.22}$$

Some steps of the solution are given

$$\begin{aligned} \frac{1}{x_1^2}\sin^2\varphi &= \frac{1}{x_2^2}\left(1-\sin^2\varphi\right) \\ \sin^2\varphi &= \frac{x_1^2}{x_1^2+x_2^2} \\ \sin\varphi &= \pm\frac{x_1}{\sqrt{x_1^2+x_2^2}} \\ B(z) &= \pm\sqrt{x_1^2+x_2^2} \end{aligned} \tag{7.23}$$

The results have to be checked using the original equation due to the unclear sign in Eq. (7.23). Hence, the sound pressure on the acoustic axis can be calculated for any distance z and any time t

$$p_A(z,t) = B(z)\sin(\omega t+\varphi) \tag{7.24}$$

Here, only the maximum of the sound pressure $p_{\max}$ is of interest, resulting in

$$p_{A\max} = |B(z)| \tag{7.25}$$

7.2.2 Back Wall Echo Curve

The ultrasound is totally reflected at the back wall coming back to the transducer. Hence, the following integrals have to be solved with the transducer being the transmitter and the receiver [2]:

$$p_{BW}(z,t) \propto 4\pi\omega Z \int_{x=0}^{\frac{D}{2}}\int_{y=0}^{\frac{D}{2}} e^{A\left(z-\sqrt{z^2+x^2}\right)^2}\frac{x\,y}{\sqrt{x^2+z^2}}\cos\left(\omega t-k\sqrt{x^2+z^2}\right)\,dx\,dy \tag{7.26}$$

with z: sound path.

7.2.3 ERS Curves

The following assumption has been made for the calculation of the ERS curves: The flat-bottomed hole oscillates with its entire area with the calculated sound pressure $p_A(z,t)$ on the acoustic axis in the distance z. Only the shortest distances (shortest time of flight) between transducer and flat-bottomed holes are considered. The double integral over the sphere area of the flat-bottomed hole as transmitter and the transducer area as receiver defines the ERS curve. Hence, the following integral has to be solved, again using the trigonometric addition theorems as described in Sect. 7.2.1:

$$p_{ers}(z,t) \propto p_A(z) \int_{x=0}^{\frac{D}{2}} \int_{y=0}^{\frac{D_{ERS}}{2}} e^{A\left(2z-\sqrt{4z^2+x^2}\right)^2} \frac{x\,y}{\sqrt{x^2+z^2}} \cos\left(\omega t - k\sqrt{x^2+4z^2}\right) dx\,dy \tag{7.27}$$

with:

- z: sound path
- D: transducer diameter
- D_{ERS}: diameter of the flat-bottomed hole
- $p_A(z)$: maximal sound pressure on the acoustic axis in the distance z

After solving the integral (7.27) for all required ERS curves these curves have to be shifted to the correct distance to the back wall echo curve. The calculation of the distances is based on the Krautkrämer book [3, pp. 97–99]. These considerations are only valid for the far field and can therefore be taken over for pulses.

On pp. 97 and 99 the Eqs. (5.4) and (5.5) are given as

$$\begin{aligned} D_r &= \frac{D_t\,z}{\pi N}\sqrt{\frac{H_r}{H_0}} \\ \frac{H_R}{H_0} &= \frac{\pi N}{2z} \end{aligned} \tag{7.28}$$

with:

- D_r: diameter of the reflector (flat-bottomed hole)
- D_t: diameter of the transducer as transmitter
- z: distance wit $z \gg N$
- H_r: echo height for the reflector
- H_R: echo height of the reference echo (back wall)
- H_0: echo height for the sound pressure on the acoustic axis at distance z
- N: near field length

The following normalization is suggested:

- $\frac{z}{N}$ = distance of reflector D,
- $\frac{H_r}{H_0}$ = amplifier gain G,
- $\frac{D_r}{D_t}$ = reflector size S.

A quotation from the Krautkrämer book [3] follows:

All the normalized values D, G, and S are dimensionless quantities and the gain G represents the ratio by which the reflector echo has to be amplified to make it equal to the reference echo. Introducing D, G, and S into the Eqs. (5.4) and (5.5) we obtain for the distant field

$$G_r = \pi \frac{S^2}{D^2}$$
$$G_R = \pi \frac{1}{2\,D}$$

But when the normalizations are inserted into Eq. (7.28) and these equations are then changed accordingly, the result is

$$D_r = \frac{D_t\,z}{\pi N}\sqrt{\frac{H_r}{H_0}}$$
$$\frac{H_R}{H_0} = \frac{\pi N}{2z} \tag{7.29}$$

In the Krautkrämer book [3], the character D is used to abbreviate the diameter for the transducer as well as abbreviation of the normalized distance of the reflector. This can be quite confusing, therefore, in the following, A is used for the normalized distance.

$$G_r = \pi^2 \frac{S^2}{A^2}$$
$$G_R = \pi \frac{1}{2\,A} \tag{7.30}$$

Comparing Eq. (7.30) with the one from the quotation it is clear that there is a misprint in the Krautkrämer book: the square of π is missing in the first equation.

When the Eq. (7.28) are written without the normalization for the size and the distance and instead of N the numerically calculated near field length N_{num} is used follows:

$$G_r = \frac{\pi^2\,D_r^2\,N_{num}^2}{z^2\,D_t^2}$$
$$G_R = \frac{\pi\,N_{num}}{2\,z} \tag{7.31}$$

Hence, the shift v_{fbh} between the back wall curve and a flat-bottomed hole curve for very large distances z is given in dB by

$$v_{fbh} = 20 \ \lg\left(\frac{G_r}{G_R}\right) \tag{7.32}$$

$$\Rightarrow \quad v_{fbh} = 20 \ \lg\left(\frac{2\,\pi\, D_r^2\, N_{num}}{D_t^2\, z}\right)$$

If necessary, D_t is replaced by $D_{eff} = 0.97\ D_t$.

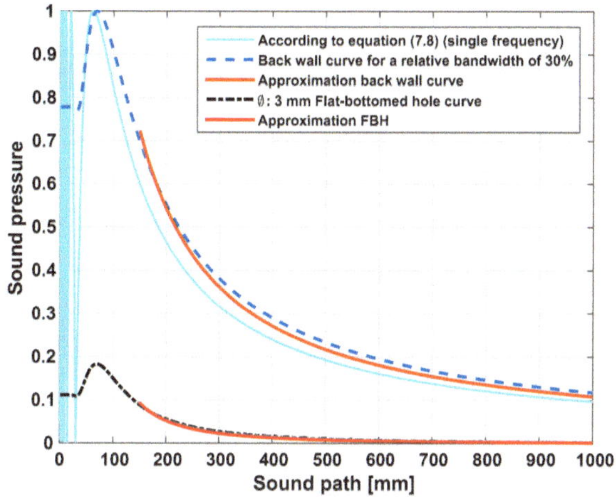

Fig. 7.9 Approximation of the back wall echo curve and the 3.1 mm ERS curve

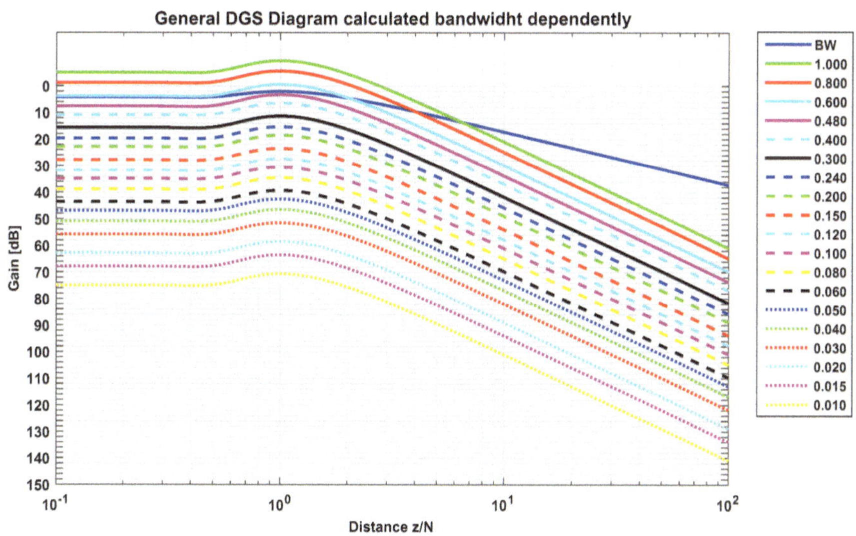

Fig. 7.10 DGS diagram calculated bandwidth dependently

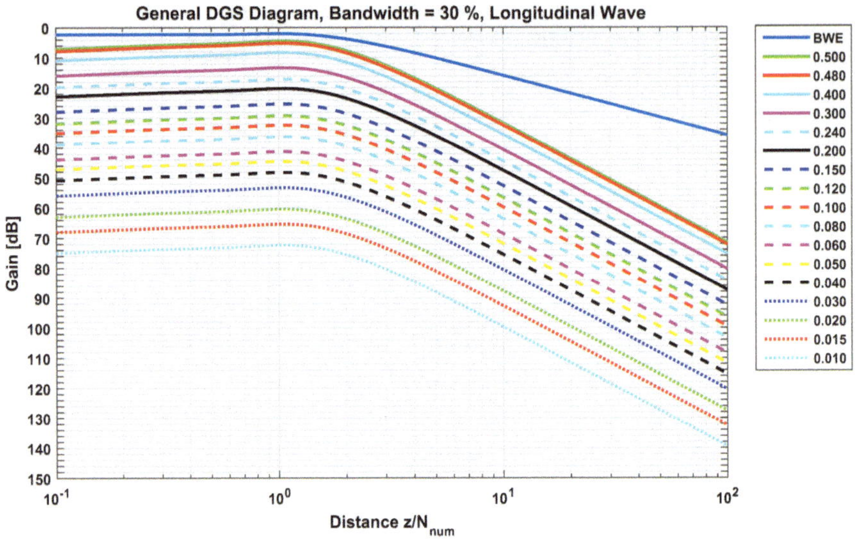

Fig. 7.11 General DGS diagram calculated for longitudinal waves with a relative bandwidth of 30 %

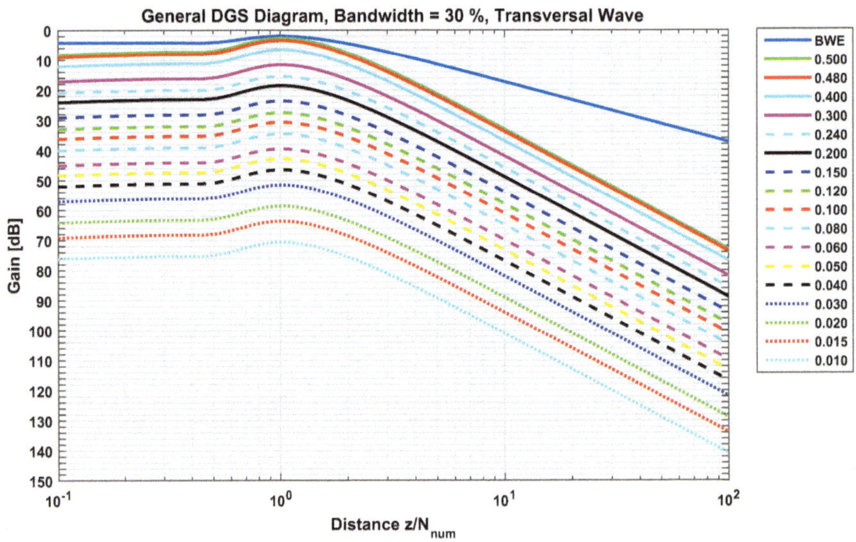

Fig. 7.12 General DGS diagram calculated for transversal waves with a relative bandwidth of 30 %

Figure 7.9 shows the good agreement of the calculated curves with the given approximation in the far field.

Surprisingly, the ERS curves for flat-bottomed holes with large diameters (large compared to the diameter of the transducer) are in certain sound path ranges above the back wall echo curve, Fig. 7.10. A back wall is working like a mirror while flat-bottomed holes generate by themselves a sound field with the maximum sound pressure at the end of their near fields. Even in the general DGS diagram, published in the EN ISO 16811:2012, the ERS curves for large (compared to the transducer diameter) flat-bottomed holes are very close together and close to the back wall echo curve as can be seen in Fig. 1.3.

In the following, equivalent reflector sizes up to half of the transducer diameter are used only with bandwidth-dependent calculated DGS diagrams. If larger flat-bottomed holes have to be evaluated a probe with a larger transducer diameter has to be applied.

Based on the introduced method bandwidth-dependent DGS diagrams have been calculated for longitudinal and for transversal waves using a relative bandwidth of 30 %. Comparing Figs. 7.11 and 7.12 shows that the DGS diagrams are different for longitudinal and for transversal waves. But for each wave mode, a general DGS diagram can be calculated for a certain bandwidth. Such a general DGS diagram is valid, for example, for a family of single element angle beam probes and for a family of phased array angle beam probes with a given bandwidth. Based on such a general DGS diagram, the special diagrams for different angles, transducer diameters, and frequencies can be derived easily. Due to this fact, it is easy to incorporate bandwidth-dependent DGS diagrams into ultrasonic instruments. It is sufficient to store the general diagrams in the instrument. The incorporation of the bandwidth-dependent DGS diagrams requires the same complexity as the incorporation of the general DGS diagram from the EN ISO 16811:2012 [4].

References

1. Boehm, R., Erhard, A., Vierke, J.: Anwendung von Modellen zur Echohöhenbewertung von Prüfköpfen mit ungewöhnlicher Schwingergeometrie, DGZfP-Berichtsband 94-CD, Plakat 44, DGZfP-Jahrestagung, Rostock (2005)
2. Kleinert, W., Oberdörfer, Y.: Calculated bandwidth dependent DGS and DAC curves for phased array sizing. In: Proceedings ECNDT, Prag. http://www.ndt.net/events/ECNDT2014/app/content/Paper/165_Kleinert.pdf (2014)
3. Krautkramer, J., Krautkramer, H.: Ultrasonic Testing of Materials. 4th Fully Revised Edition Translation of the 5th Revised German Edition. Springer, Berlin (1990)
4. Ultrasonic testing - Sensitivity and range setting (ISO 16811:2012); German version EN ISO 16811:2014

Chapter 8
Applying Bandwidth-Dependent DGS Diagrams

Abstract Measurements were taken using ten units each of 2 and 4 MHz new phased array angle beam probes. Special test blocks with planar back walls and flat-bottomed holes in different depths were used for five different angles of incidence. All measurements were taken manually by different operators. The minimum sound path was 5 mm. The DGS evaluation of this huge amount of measurement values shows high accuracy over the entire range of sound paths. With these bandwidth-dependent DGS diagrams, the restriction in the EN ISO 16811:2012 [1] to use only sound paths larger than 0.7 near field length can be neglected.

The bandwidth-dependent DGS diagrams for a relative bandwidth of 30 % were used to evaluate measurements with *true*DGS® phased array angle beam probes with frequencies of 2 and 4 MHz. When using phased array angle beam probes all DGS-related parameters are changed with the steering angle applied. Adapting these changing parameters can be done easily by deriving the required special diagram from the general DGS diagram.

First, the DGS accuracy of the bandwidth-dependent DGS diagram is tested using a 4 MHz phased array angle beam probe with a steering angle of 60°. The result is compared to the evaluation using the general DGS diagram published in the EN ISO 16811:2012, refer to Fig. 7.4.

The result of the evaluation using the bandwidth-dependent DGS diagram is illustrated in Fig. 8.1. The evaluation shows an excellent agreement of the measurement values with the bandwidth-dependent DGS diagram in the entire range of sound paths used for measurements. The comparison of this result with the result using the general DGS diagram published in the EN ISO 16811:2012 [1] proves the advantage of applying bandwidth-dependent DGS diagrams. The DGS accuracy achieved is shown as well in Fig. 8.2 evaluating measurements with a 2 MHZ phased array angle beam probe at a steering angle of 65°.

The restriction to sound paths >0.7 N as requested by the EN ISO 16811:2012 [1] can be neglected when bandwidth-dependent DGS diagrams are used.

W. Kleinert, *Defect Sizing Using Non-destructive Ultrasonic Testing*,
DOI 10.1007/978-3-319-32836-2_8

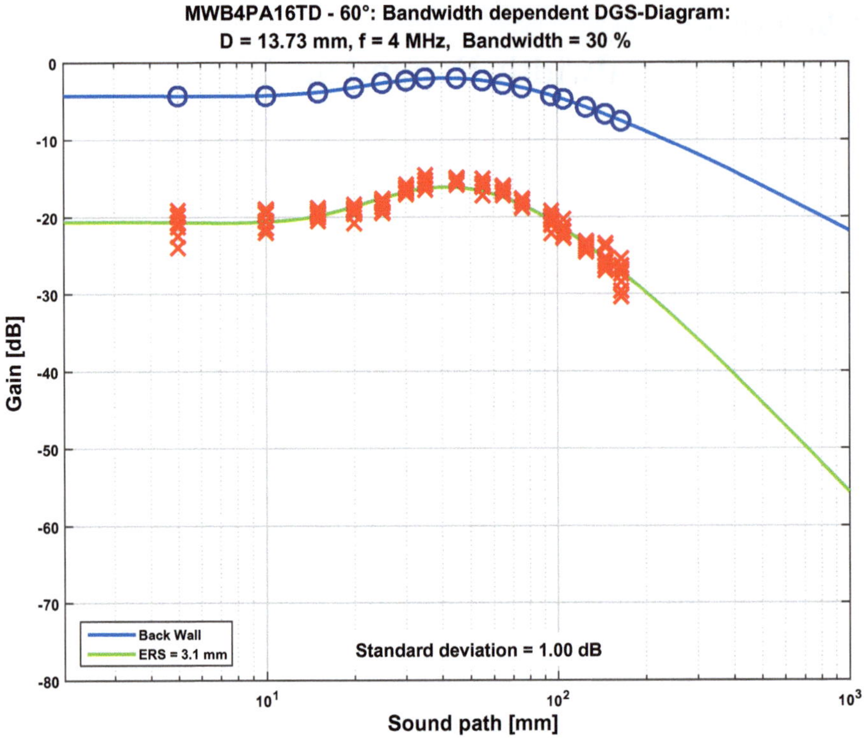

Fig. 8.1 Evaluation based on a bandwidth-dependent DGS diagram covering the entire range of sound paths

8.1 Results Using Phased Array Angle Beam Probes

The measurements were taken by different ultrasonic operators using ten prototypes of 2 and 4 MHz *true*DGS® phased array angle beam probes. For each measured flat-bottomed hole, a back wall was measured at the same sound path as the flat-bottomed hole to avoid any influences of sound attenuation in the used test blocks. All measurements have been taken manually. The results of the DGS evaluation are shown in Tables 8.1 and 8.2. All results show a high DGS accuracy in the entire range of sound paths used [2].

DGS diagrams with the evaluation for both 2 and 4 MHz *true*DGS® phased array angle beam probes can be found in the Appendix.

Figure 8.3 shows a tool for applying the bandwidth-dependent general DGS diagram. In this case, a DGS evaluation of a 2 MHz *true*DGS® phased array angle beam probe using just one single reference for all other measurements is illustrated.

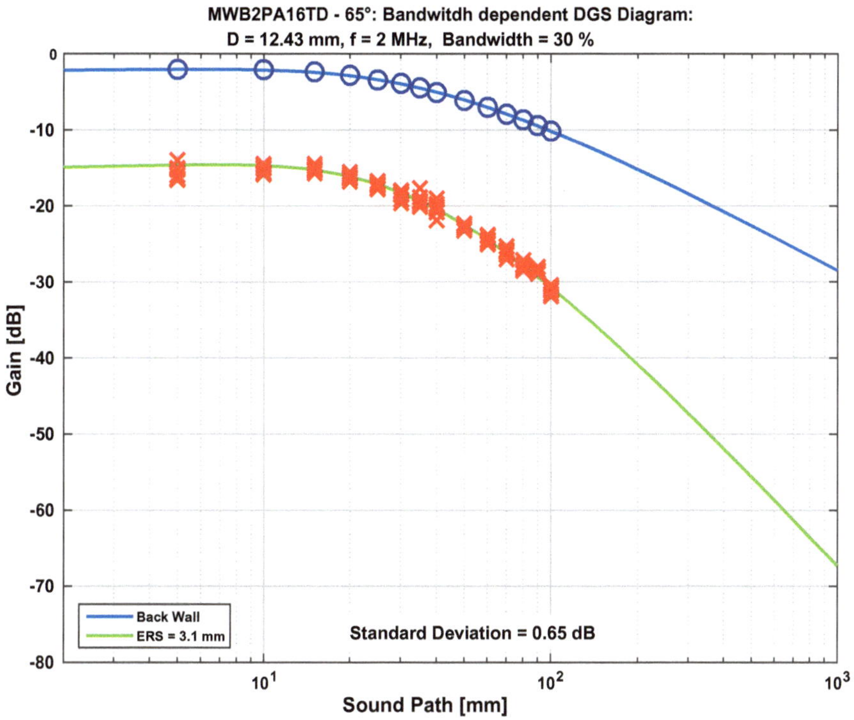

Fig. 8.2 Evaluation of measurements taken with a *true*DGS® 2 MHz phased array angle beam probe with a steering angle of 65°

Table 8.1 Result of the evaluation of measurements taken with the 2 MHz *true*DGS® phased array angle beam probe

Steering angle (°)	Standard deviation (dB)
45	1.30
53	0.87
60	0.97
65	0.65
70	1.04

Table 8.2 Result of the evaluation of measurements taken with the 4 MHz *true*DGS® phased array angle beam probe

Steering angle (°)	Standard deviation (dB)
45	1.45
53	1.05
60	1.02
65	1.38
70	1.03

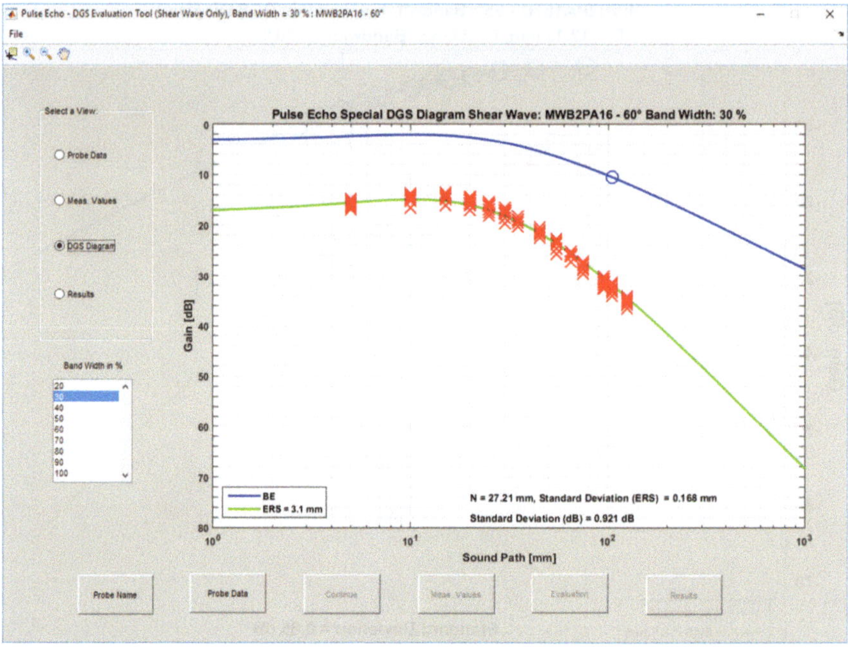

Fig. 8.3 DGS evaluation using one single reference echo

References

1. Ultrasonic testing—Sensitivity and range setting (ISO 16811:2012); German version EN ISO 16811:2014
2. Kleinert W., Oberdörfer Y.: Calculated Bandwidth Dependent DGS and DAC Curves for Phased Array Sizing Proceedings ECNDT, Prag (2014). http://www.ndt.net/events/ECNDT2014/app/content/Paper/165_Kleinert.pdf

Chapter 9
Bandwidth-Dependent DAC Curves

Abstract Defect sizing using Distance Amplitude Correction (DAC) curves gets very time consuming when using phased array angle beam probes. For each incidence angle applied during testing a DAC/TCG curve has to be recorded. By changing the phasing angle important probe parameters such as near field length, delay line length and sensitivity are changed as well, resulting in a change of the DAC/TCG curve. DAC curves can be calculated using the new probe technology. It is sufficient to just measure one reference echo, e.g., using a calibration standard. The DAC curves for all angles can be calculated. Alternatively a DAC curve can be recorded using one single angle of the phased array angle beam probe. By recording the curve for one single angle the material characteristics of the reference block such as the sound attenuation are considered automatically. The DAC curve for all other angles are calculated with high accuracy. With the new probe technology recording DAC curves for phased array angle beam probes is as easy as recording a DAC curve with a single element angle beam probe.

In the USA, different organizations such as ASME, ASTM, and AWS publish standards requesting the recording of Distance Amplitude Correction (DAC) curves. For recording of DAC curves expensive reference blocks are necessary. Defect sizing, particularly when using phased array angle beam probes, requires recording of DAC curves for each angle to be applied for testing which is very time consuming. When using the *true*DGS® technology DAC curves can be calculated with high precision [3].

9.1 Calculating Bandwidth-Dependent DAC Curves

The sound pressure on the acoustic axis generated by a circular reflector or transducer decreases depending on the distance z by:

$$\frac{1}{z}$$

For side-drilled holes, the decrease is given by

W. Kleinert, *Defect Sizing Using Non-destructive Ultrasonic Testing*,
DOI 10.1007/978-3-319-32836-2_9

$$\frac{1}{\sqrt{z}}$$

While calculating the DAC curves, it has to be considered that the way to the reflector is following the decrease based on the circular transducer. On the way back from the side-drilled hole to the transducer the decrease is following the distance law for side-drilled holes. DAC curves for side-drilled holes are only dependent on the probe used (near field length and delay line length). The diameter of the side-drilled hole changes the sensitivity only. Based on these facts a general DAC curve can be calculated if *true*DGS® phased array angle beam probes are used (Fig. 9.1).

The back wall echo curve is calculated using Eq. (7.26). The DAC curve is given by

$$p_{sdh}(z,t) \propto p_A(z) \int_0^{\frac{D}{2}} e^{A\left(2z-\sqrt{4z^2+x^2}\right)^2} \frac{x}{\sqrt[4]{x^2+z^2}} \cos\left(\omega t - k\sqrt{x^2+4z^2}\right) dx \tag{9.1}$$

with:

- z: sound path
- $p_A(z)$: max. sound pressure on the acoustic axis at distance z
- D: transducer diameter
- ω: angular frequency
- k: angular wavenumber
- p_{sdh}: resulting sound pressure from the side-drilled hole.

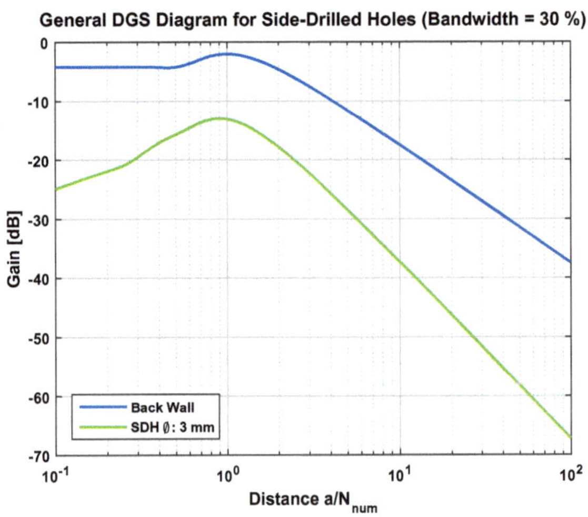

Fig. 9.1 General DGS diagram for side-drilled holes

Three different ways to use the bandwidth-dependent calculated DAC will be discussed:

- using a reference echo from a calibration standard
- using the measured result from one single side-drilled hole as reference
- recording a DAC curve for one single angle using the Least Squares Method.

For the first case, the distance between the back wall curve and the DAC curve has to be derived. All following considerations apply to the distant field only. The sound pressure p on the acoustic axis of a circular transducer can be calculated by [1]

$$p \approx p_0 \frac{\pi \, D_t^2}{4 \, \lambda \, z} \tag{9.2}$$

The sound pressure of a side-drilled hole follows the distance law in the far field [2]:

$$\frac{\sqrt{D_z}}{\sqrt{z}} \tag{9.3}$$

with D_z: diameter of the side-drilled hole.

Hence, the sound pressure p_{sdh} of a side-drilled hole in the sound field of a circular transducer is described by the product of the Eq. (9.2). Equation (9.2) can be considered as the sound going forward to the reflector and as the sound traveling back to the transducer. That means, in the product (9.4), z has to be understood as the sound path to the reflector.

$$p_{sdh} = p_0 \frac{\pi \; D_t^2 \sqrt{D_z}}{4 \, \lambda \, z^{\frac{3}{2}}} \tag{9.4}$$

Figure 9.2 illustrates the good approximation of the sound pressure of a side-drilled hole in the sound field of a circular transducer in the distant field by Eq. (9.4).

The sound pressure of a planar back wall in the far field is given by [1]

$$p_R = p_0 \frac{\pi \, N_{num}}{z} \tag{9.5}$$

with N_{num} near field length of the circular transducer calculated numerically.

Note: z stands here for the sound path! The ratio of P_{sdh} and p_R describes the ratio of the amplitudes H_R of the back wall echo and the echo of the side-drilled hole H_{sdh}:

$$\frac{H_{sdh}}{H_R} = \frac{D_t^2 \sqrt{D_z}}{4 \, \lambda \, N_{num} \sqrt{z}} = \frac{N \sqrt{D_z}}{N_{num} \sqrt{z}} \tag{9.6}$$

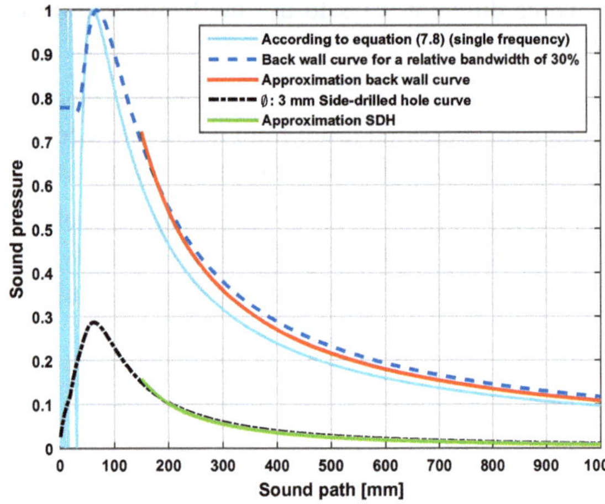

Fig. 9.2 Approximation of the sound pressure in the far field

with D_z being the diameter of the side-drilled hole and z the sound path. With this the shift v_{sdh} between the back wall echo curve and the curve describing the sound pressure of a side-drilled hole with diameter D_z in the distant field results in

$$v_{sdh} = 20 \ \lg \left(\frac{N \ \sqrt{D_z}}{N_{num} \ \sqrt{z}} \right) \tag{9.7}$$

If necessary, D_t is replaced by $D_{eff} = 0.97 \ D_t$.

9.2 Applying the Bandwidth-Dependent DAC Curves

As already mentioned, there are three different ways to use the bandwidth-dependent calculated DAC

- using a reference echo from a calibration standard
- using the measured result from one single side-drilled hole as reference
- record a DAC curve for one single angle using the least squares method

All three cases will be discussed in the following.

9.2.1 Using a Reference Echo from a Calibration Standard

This method is very similar to the method described in Chap. 2. A reference echo is taken from a calibration standard, here calibration standard K1 according to DIN EN 2400:2012 is used. The reference is taken from the circular arc of the K1 using a predefined angle. Since this is not a planar back wall a correction value ΔV_{K1} has to be considered. These correction values are published with each *true*DGS® phased array angle beam probe for each possible angle. As discussed in Chap. 2 the sound attenuation has to be known or measured. If the coupling surface qualities are different between the calibration standard and the test piece, the transfer correction value has to be taken into account as well.

First, a DAC curve for the 3 mm side-drilled hole is calculated together with the back wall echo curve as reference. In addition the DGS curve for the 3 mm side-drilled hole is adapted to the known sound attenuation of 9.5 dB/m. Figure 9.3 shows this DGS diagram for a 3 mm side-drilled hole. Let $G_{K1} = 3.0\,\text{dB}$ be the gain setting to have the echo from the circular arc of the calibration standard K1 at 80 % screen height. The amplitude correction value ΔV_{K1} is given as 4.4 dB. Let the transfer correction value be zero. With this, the reference gain is

$$G_R = G_{K1} + \Delta V_{K1} = 3.0\text{ dB} + 4.4\text{ dB} = 7.4\text{ dB}$$

The reference echo at a sound path of 100 mm is marked on the back wall echo curve in the respective DGS diagram. The SDH curve is adapted for the sound attenuation (dashed line in Fig. 9.4). The dB-difference ΔG between the reference echo on the back wall echo curve and the maximum of the adapted line is taken from the diagram, here $\Delta G = 10.2\,\text{dB}$. The gain setting G of the ultrasonic instrument is set to:

$$G = G_R + \Delta G = 7.4\text{ dB} + 10.2\text{ dB} = 17.6\text{ dB}$$

Now, the curve that is adapted for the sound attenuation is linearized so that the maximum corresponds to 80 % screen height. With this gain setting each side-drilled hole (same diameter preconditioned) at different sound paths generates an

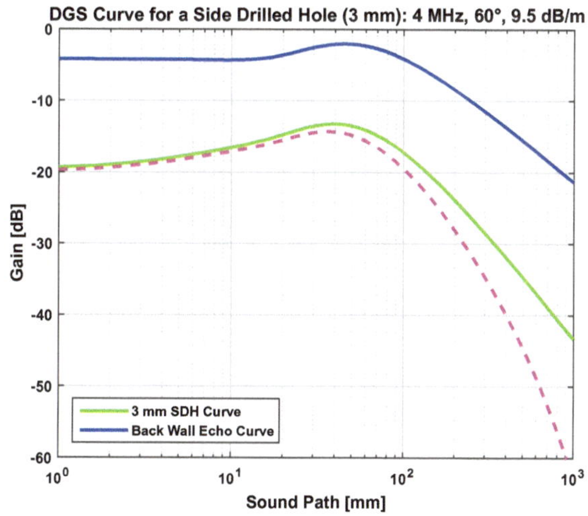

Fig. 9.3 DGS curve for a side-drilled hole

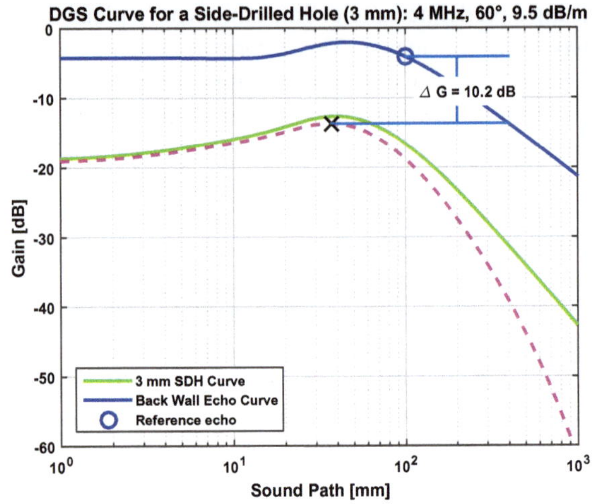

Fig. 9.4 DGS diagram for a 3 mm SDH with reference echo and ΔG marked

echo amplitude just reaching the calculated and displayed curve. Figure 9.5 shows an example of such a curve. But not only the curve for the angle used to record the reference echo can be calculated. With *true*DGS® phased array angle beam probes all curves for all other angles can be calculated as well. In order to validate the accuracy of the calculated curve the echoes of all seven side-drilled holes were recorded using the calculated gain setting. In Table 9.1, the deviations in dB between the calculated curve and the measurement values are listed and they show the very good agreement with the theory, refer as well to Fig. 9.6.

The curves for the other angles of the *true*DGS® phased array angle beam probe can be calculated easily since all needed values are documented with these probes.

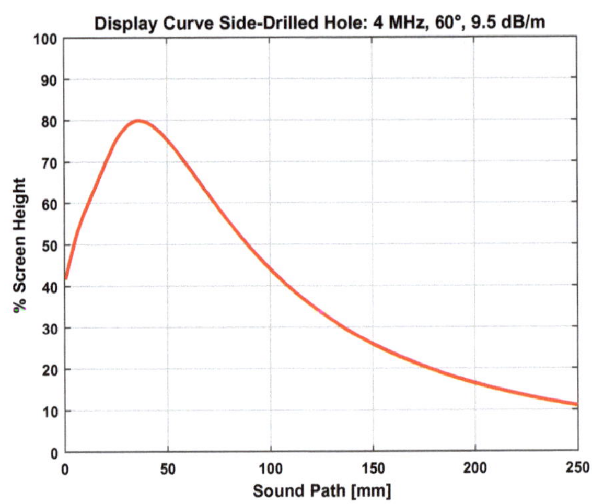

Fig. 9.5 Calculated display curve

Table 9.1 Deviations between calculated curve and measurement values using a reference echo from the calibration block K1

Sound path (mm)	Deviation (dB)
19.5	1.57
58.0	0.12
97.0	0.36
138.0	0.70
175.0	–0.40
216.0	–1.06
254.0	0.59

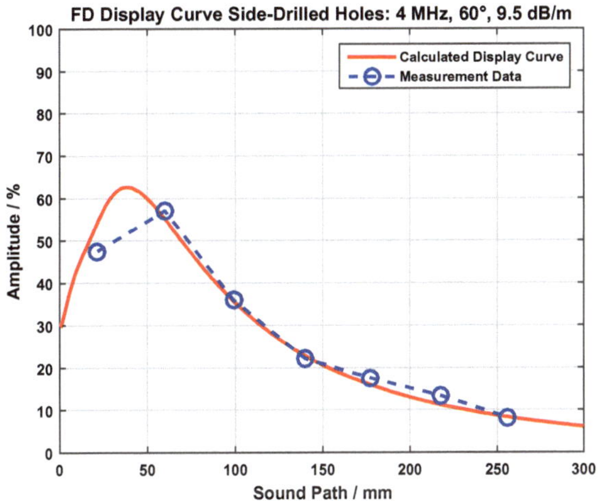

Fig. 9.6 Calculated display curve with measurement results

For each angle a different sensitivity as well as a different delay line and a different near field length need to be taken into account.

9.2.2 *Using One Single Side-Drilled Hole as Reference*

Alternatively, one single side-drilled hole can be used as reference. The curve for a side-drilled hole with a given diameter is dependent only on the probe sensitivity and the delay line length. The delay line length is known from the data sheet of the probe and the sensitivity is determined by the side-drilled hole used as reference. It is assumed that the side-drilled hole taken for reference is hitting the calculated display curve. The precondition is that the sound attenuation in the test block is known. On the left side of Fig. 9.7, the calculated display curve is shown. The reference echo from the side-drilled hole at a sound path of 48.5 mm is marked. All other side-drilled holes in the test block were measured as well to validate the calculated curve. On the right side of Fig. 9.7, the display curve is shown together with the measurements of

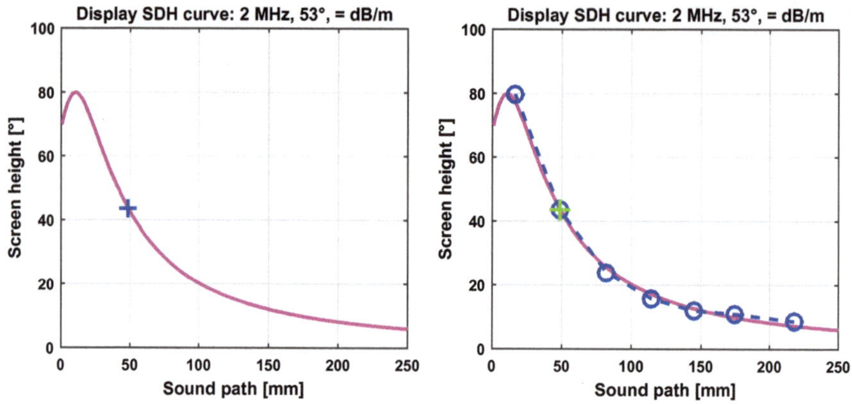

Fig. 9.7 Using one single SDH as reference and validation applying the rest of the side-drilled holes in the test block

all side-drilled holes available in the reference block. Of course the display curves for the other angles of the *true*DGS® phased array angle beam probe can be calculated in this case as well. The deviations of the measured echoes of all side-drilled holes are listed in Table 9.2. All manually measured values are in a good agreement with the calculated curve.

9.2.3 Recording a DAC Curve for One Single Angle

Cases where reference echoes are used are heavily dependent on the accuracy of the measurement of the reference echo. Any deviation by taking the reference echo will as well be seen in the evaluation of all other measurements. Additionally, the sound attenuation has to be known or to be measured.

Alternatively, all side-drilled holes available in a reference block are recorded for one single angle of the used phased array angle beam probe. The *Least Squares*

Table 9.2 Deviations between calculated curve and measurement values

Sound path (mm)	Deviation (dB)
16.5	–0.27
48.5	**0**
82.0	0.55
114.0	0.73
145.0	0.46
175.0	–0.90
218.0	–1.50

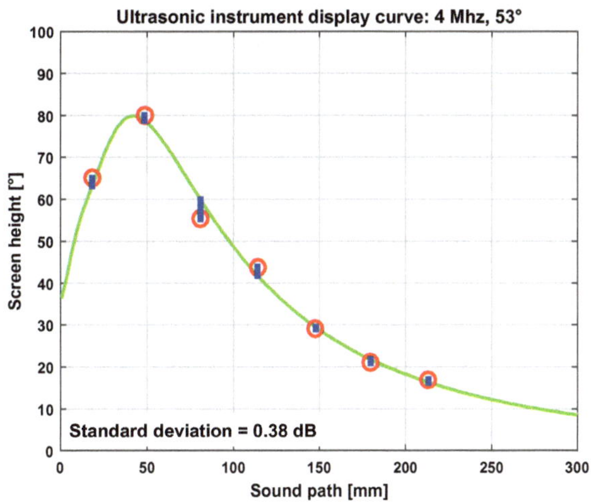

Fig. 9.8 Display curve with measurement values and minimized distances between measurements and calculated curve

Method is applied to get a best fit of the measurement values to the calculated curve. The *Least Squares Method* is used to determine the sensitivity of the probe in relation to the side-drilled holes and in addition will determine the sound attenuation in the reference block. Since the reference block is mostly produced from the same material as the test piece, the sound attenuation in the test piece is then known as well.

The *Least Squares Method* is used to minimize the distances between the measurement values and the calculated curve (Fig. 9.8). Two parameters are taken into account to minimize the sum of the squared distances between measurements and curve

- probe sensitivity in relation to the side-drilled hole
- sound attenuation in the test block

For applying the *Least Squares Method* the following variables are used:

- n: amount of measurement values available
- s_i: measured sound paths
- r_i: dB values on the calculated DGS curve for side-drilled holes at the measured sound paths
- m_i: measured gain values in dB
- κ: sound attenuation to be derived
- v: vertical shift of the DAC curve for side-drilled holes based on the sensitivity of the probe used

The sum F of the squared distances between the measurements and the calculated curve is given by

$$F = \sum_{i=1}^{n} [(r_i - 2\kappa s_i) - m_i + v]^2 \tag{9.8}$$

The term $r_i - 2\kappa s_i$ describes the DGS curve for side-drilled holes after being corrected for the sound attenuation in the reference block.

The two partial deviations of Eq. 9.8 are derived and set to zero:

$$\frac{\partial F}{\partial v} = 0 \text{ and } \frac{\partial F}{\partial \kappa} = 0 \tag{9.9}$$

Hence, the following systems of equations is given:

$$\begin{pmatrix} 2\sum_{i=1}^{n} s_i & -n \\ 2\sum_{i=1}^{n} s_i^2 - \sum_{i=1}^{n} s_i \end{pmatrix} \begin{pmatrix} \kappa \\ v \end{pmatrix} = \begin{pmatrix} \sum_{i=1}^{n} (r_i - m_i) \\ \sum_{i=1}^{n} s_i (r_i - m_i) \end{pmatrix} \tag{9.10}$$

resulting in the best fit of the measurements to the calculated curve considering the sound attenuation.

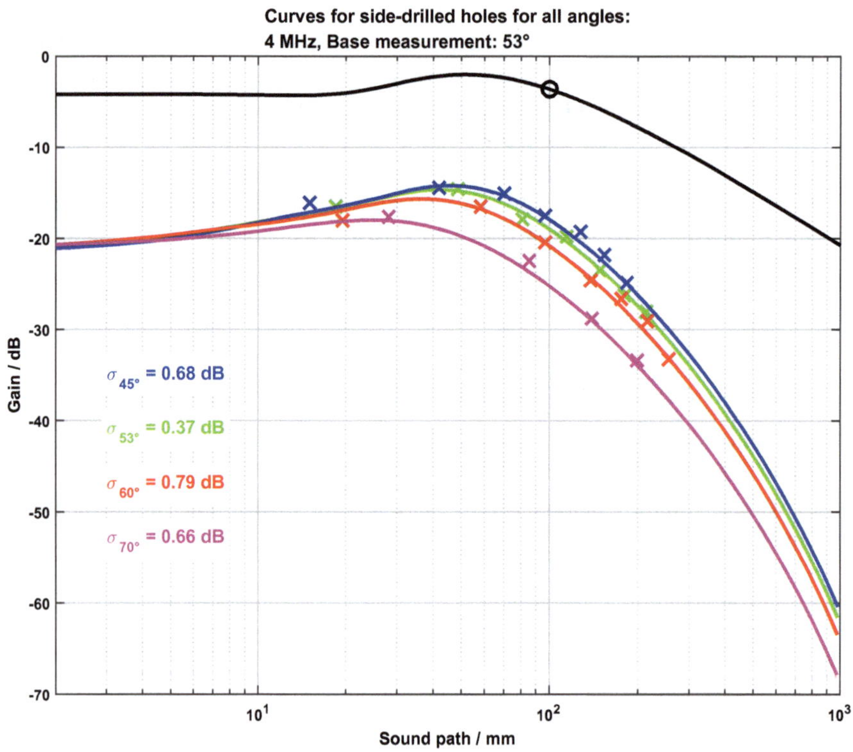

Fig. 9.9 DGS curves for side-drilled holes recorded using four different angles including the standard deviations between measurements and calculated curves

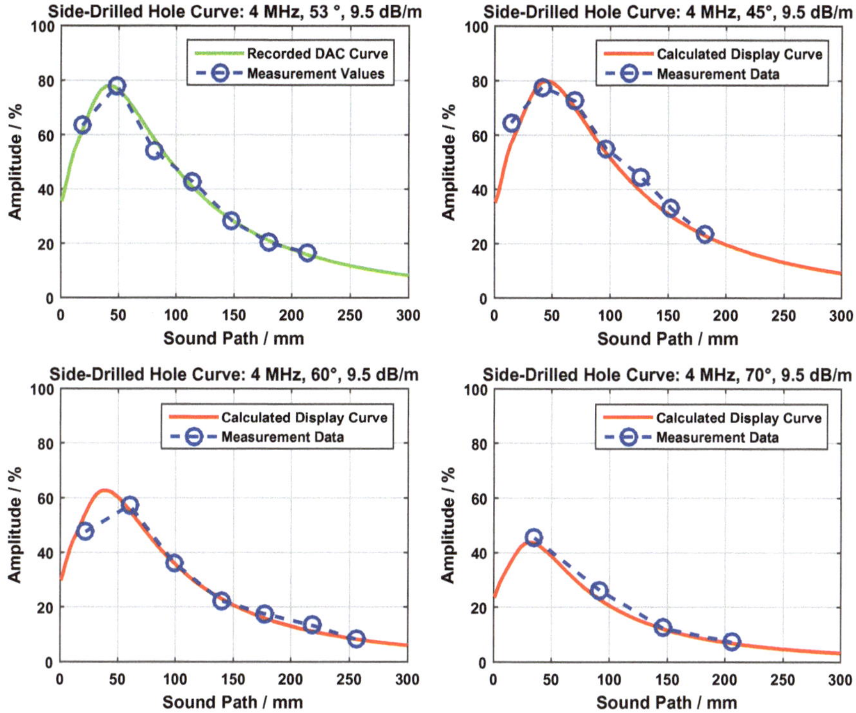

Fig. 9.10 Validation of calculated curves for the other three angles

Using a 4 MHz *true*DGS® phased array angle beam probe seven side-drilled holes (diameter: 3 mm) were recorded for an angle of 53°. The DGS curves for side-drilled holes were calculated for the three remaining angles under consideration of the differences depending on the steering angle. For all other angles the side-drilled holes were recorded for additional validation. Figure 9.9 shows the result including the standard deviations reached for each angle.

Again, for validation all side-drilled holes were measured for 45°, 60° and 70° and compared to the calculated curves. Figure 9.10 shows the results and the good agreement between measurement and calculated curves. In all subfigures of Fig. 9.10, the display curves are shown for the same gain setting. The deviations between measurement data and calculated curves achieved with a 4 MHz *true*DGS® phased array angle beam probe are listed in Tables 9.3 and 9.4. The corresponding data for the 2 MHz phase array angle beam probe is listed in Tables 9.5 and 9.6.

Instead of the DAC curve on the ultrasonic instrument, a function called Time Corrected Gain (TCG) can be used. This function ensures that all echoes which would reach the DAC curve are set to 80 % screen height of the instrument. Figure 9.11 illustrates the TCG function.

Table 9.3 Deviations between calculated curve and measurements taken with a 4 MHz *true*DGS® phased array angle beam probe with steering angles of 53° and 45°

Frequency	Sound path	Deviation @ 53°	Sound path	Deviation @ 45°
4 MHz	18.5	0.23	15.0	1.09
4 MHz	48.5	0.12	42.0	–0.19
4 MHz	81.0	–0.68	70.0	0.31
4 MHz	114.0	0.44	97.0	0.24
4 MHz	148.0	–0.13	127.0	1.15
4 MHz	180.0	–0.33	153.0	0.78
4 MHz	213.5	0.35	183.0	0.16

Table 9.4 Deviations between calculated curve and measurements taken with a 4 MHz *true*DGS® phased array angle beam probe with steering angles of 60° and 70°

Frequency	Sound path	Deviation @ 60°	Sound path	Deviation @ 70°
4 MHz	19.5	–1.14	28.0	0.39
4 MHz	58.0	0.31	85.0	1.19
4 MHz	79.0	0.07	139.0	0.18
4 MHz	138.0	–0.27	199.0	0.46
4 MHz	175.0	0.83		
4 MHz	216.0	1.49		
4 MHz	254.0	–0.15		

Table 9.5 Deviations between calculated curve and measurements taken with a 2 MHz *true*DGS® phased array angle beam probe with steering angles of 53° and 45°

Frequency	Sound path	Deviation @ 53°	Sound path	Deviation @ 45°
2 MHz	16.5	0.63	15.0	1.34
2 MHz	48.5	–0.09	42.5	1.09
2 MHz	82.0	–0.75	70.0	0.33
2 MHz	114.0	–0.96	97.0	–0.34
2 MHz	145.0	–0.71	127.0	0.11
2 MHz	175.0	0.65	153.0	–1.02
2 MHz	218.0	1.24		

9.2.4 *Pros and Cons*

Whenever a reference echo is used, the accuracy of the entire evaluation depends on the accuracy of the measurement of the reference echo. That the sound attenuation has to be known or to be measured is another disadvantage. When the reference echo

Table 9.6 Deviations between calculated curve and measurements taken with a 2 MHz *true*DGS® phased array angle beam probe with steering angles of 60° and 70°

Frequency	Sound path	Deviation @ 60°	Sound path	Deviation @ 70°
2 MHz	19.0	0.90	25.0	1.83
2 MHz	59.0	–0.36	80.0	0.67
2 MHz	96.0	–1.21	150.0	0.95
2 MHz	137.0	–0.61		
2 MHz	173.0	–0.37		
2 MHz	210.0	–0.70		

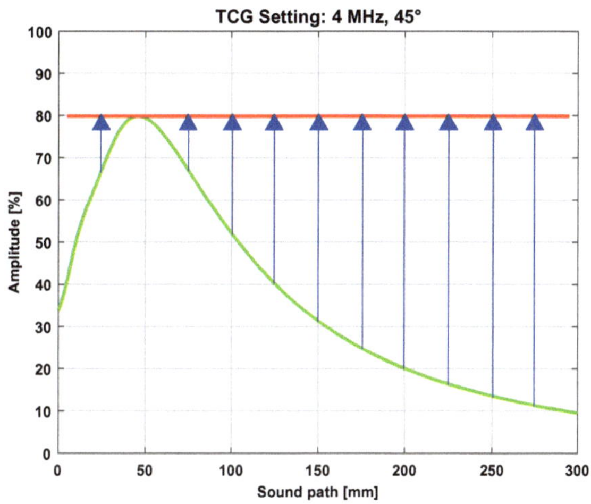

Fig. 9.11 Alternatively to the DAC display curve time corrected gain can be used

is taken from the circular arc of the calibration standard, the amplitude correction value ΔV_K has to be taken into account.

The recording of a DAC curve with one single steering angle of a *true*DGS® phased array angle beam probe by applying the *Least Squares Method* measures automatically the sound attenuation in the test block. The issue, that DAC curves have to be recorded for each angle applied for testing, is solved when *true*DGS® phased array angle beam probes are used since the DAC curve for all other angles can be calculated with high accuracy.

By this approach defect sizing using phased array probes is as easy as using a single element angle beam probe.

References

1. Krautkramer, J., Krautkramer, H.: Ultrasonic testing of materials. In: 4th Fully Revised Edition Translation of the 5th Revised German Edition. Springer, Heidelberg GmbH (1990)
2. Vierke, J.: Empfindlichkeitseinstellung und Echohöhenbewertung von Prüfköpfen mit schmalen rechteckigen Schwingern, Diplomarbeit (2004)
3. Kleinert, W., Oberdörfer, Y.: Calculated bandwidth dependent DGS and DAC curves for phased array sizing. In: Proceedings ECNDT 2014 Prag. http://www.ndt.net/events/ECNDT2014/app/content/Paper/165_Kleinert.pdf

Chapter 10
Convert SDH into FBH and Vice Versa

Abstract The new probe technology is applied to calculate for a given sound path and a given side-drilled hole diameter an equivalent flat-bottomed hole diameter. The term equivalent diameters is used when both holes are generating the same amplitude at the given sound path. This method can be applied vice versa calculating a side-drilled hole from a given flat-bottomed hole.

The distance from the back wall echo curve to the curve of a specific flat-bottomed hole has been calculated using Eq. (7.32). The distance between the back wall curve to a curve of a side-drilled hole was determined as well, Eq. (9.8).

Equating these two formulas yields

$$\frac{2\,\pi\,D_r^2\,N_{num}}{z\,D_s^2} = \frac{N\sqrt{D_z}}{N_{num}\,\sqrt{z}} \tag{10.1}$$

When this equation is fulfilled, the flat-bottomed hole with the diameter D_r and the side-drilled hole with the diameter D_z will generate the same echo height if both are measured at a sound path of z as long as z is in the distant field.

The diameter of the equivalent flat-bottomed hole for a given side-drilled hole is derived by dissolving Eq. (10.1) using Eq. (7.13):

$$D_r = \frac{D_s^2}{2\,\sqrt{\lambda}\,N_{num}}\sqrt{\frac{1}{2\,\pi}\,\sqrt{z\,D_z}} \tag{10.2}$$

$$\Rightarrow\quad D_r = \frac{N}{N_{num}}\sqrt{\frac{2\lambda}{\pi}\,\sqrt{z\,D_z}}$$

These two Eqs. (10.1) and (10.2) are only valid if *true*DGS® probes are applied.

W. Kleinert, *Defect Sizing Using Non-destructive Ultrasonic Testing*,
DOI 10.1007/978-3-319-32836-2_10

Being able to calculate DGS curves for flat-bottomed holes and to calculate DAC curves for side-drilled holes [3] allows to determine equivalent diameters for each sound path. Diameters are called equivalent when the flat-bottomed hole and the side-drilled hole generate the same echo height at the given sound path.

First, the determination of a flat-bottomed hole for a given side-drilled hole is discussed, Fig. 10.1. A DGS diagram including the back wall curve and the curve for the given diameter of a side-drilled hole is established. The curve for the flat-bottomed hole with the same diameter is added to this diagram. The dB-difference between the two curves for the flat-bottomed hole and the side-drilled hole is derived at the desired sound path. Knowing this value yields the required total shift of the curve for the flat-bottomed hole from the back wall echo curve in the far field. Using Eq. (7.32) results in the required diameter of the equivalent flat-bottomed hole. For the sake of completeness the DGS curve for this diameter is added to the DGS diagram.

The other direction, converting a given flat-bottomed hole into a equivalent side-drilled hole, is done accordingly, Fig. 10.2. In this case it is not necessary to recalculate the shifted curve for the side-drilled hole because the shape of these curves is independent of the diameter of the side-drilled hole. In the following, the method described will be refered to as *true*DGS® method.

With the skill of converting flat-bottomed holes into equivalent side-drilled holes and vice versa any side-drilled hole can be used as a reference echo for a DGS evaluation or for the calculation of DAC curves.

Another formula to convert the diameter of a side-drilled hole into the diameter of an equivalent flat-bottomed hole can be found in literature, e.g., in DIN EN 583-2:2001 [1]

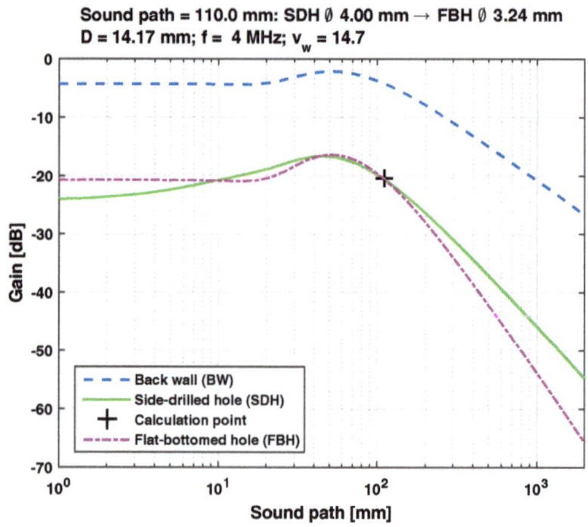

Fig. 10.1 Converting a given side-drilled hole into a flat-bottomed hole

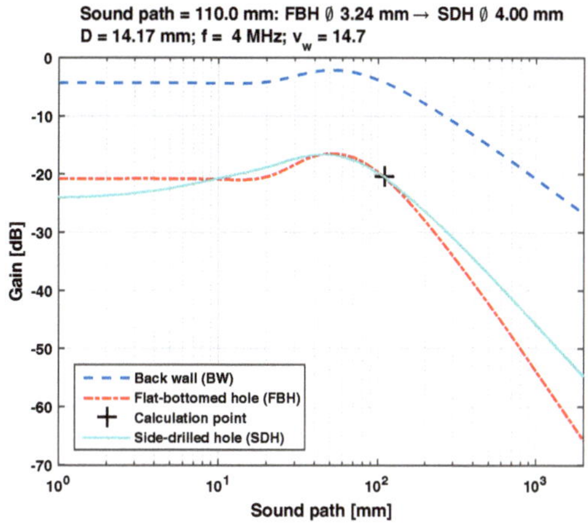

Fig. 10.2 Converting a given flat-bottomed hole into a side-drilled hole

$$D_r = \sqrt{\frac{\sqrt{2}}{\pi}\,\lambda\,\sqrt{z\,D_z}} \tag{10.3}$$

with the limitations of $D_z \geq 1.5\,\lambda$ and $z \geq 1.5\,N$. This formula is no longer present in the successor (EN ISO 16811:2012 [2]) of DIN EN 583-2:2001.

Of course at the time when this formula was developed *true*DGS® probes could not be considered.

Table 10.1 shows that Eq. (10.2) is in good agreement with the *true*DGS® method already from two near-field lengths onwards.

When looking at the differences of these three methods to calculate equivalent diameters of side-drilled and flat-bottomed holes it has to be considered, that doubling the diameter of a side-drilled hole results in a dB-difference of just 3 dB. In

Table 10.1 Comparison of all methods described to convert a side-drilled hole with diameter 4 mm into an equivalent flat-bottomed hole using a 4 MHz *true*DGS® probe

Distance	*true*DGS® method	According to Eq. (10.2)	According to Eq. (10.3)
$N_{num}/2$	3.36	2.44	2.30
N_{num}	3.15	2.77	2.61
$2\,N_{num}$	3.35	3.20	3.02
$3\,N_{num}$	3.59	3.51	3.31
$4\,N_{num}$	3.81	3.76	3.54
$5\,N_{num}$	4.00	3.97	3.74
...	...	...	...
$10\,N_{num}$	4.70	4.69	4.42
$20\,N_{num}$	5.56	5.56	5.24

Table 10.2 Comparison of all methods described to convert a side-drilled hole with diameter 4 mm into an equivalent flat-bottomed hole using a 2 MHz *true*DGS® probe

Distance	*true*DGS® method	According to Eq. (10.2)	According to Eq. (10.3)
$N_{num}/2$	3.58	2.87	2.84
N_{num}	3.55	3.18	3.14
2 N_{num}	3.77	3.61	3.56
3 N_{num}	4.01	3.92	3.87
4 N_{num}	4.24	4.18	4.12
5 N_{num}	4.44	4.39	4.34
...	...	...	...
10 N_{num}	5.18	5.16	5.10
20 N_{num}	6.11	6.10	6.02

addition these differences significantly depend on the frequency as is demonstrated in Table 10.2 showing results using a 2 MHz probe.

10.1 SDH or FBH?

The main difference between side-drilled holes and flat-bottomed holes is in the behavior in the far field, refer to Table 10.3. These differences have to be considered when diameters are given as limits for a certain ultrasonic test usually based on fracture mechanics.

Besides these different characteristics it has been shown that for both types of holes DGS and DAC curves can be calculated when *true*DGS® angle beam probes are applied. For both types general DGS diagrams can be developed for a given bandwidth. The main advantage of this technology is that when using *true*DGS® phased array angle beam probes it is sufficient to record a reference for one single angle. The DGS and DAC curves can then be calculated for all other angles. This is particularly important when DAC curves have to be recorded using phased array angle beam probes. Instead of recording DAC curves for all angles to be applied when testing ultrasonically it is sufficient to record a DAC curve for one single angle

Table 10.3 Characteristics of side-drilled holes (SDH) and flat-bottomed holes (FBH) in the distant field

	SDH	FBH
	Δ (dB)	Δ (dB)
Twice the distance	9	12
Twice the diameter	3	12

only. With the recording for one single angle the material characteristics are taken into account and the DAC curves for all other angles can be calculated with high accuracy.

References

1. DIN EN 583-2:2001 Zerstörungsfreie Prüfung-Ultraschallprüfung, Teil 2: Empfindlichkeits—und Entfernungsjustierung
2. Ultrasonic testing—Sensitivity and range setting (ISO 16811:2012); German version EN ISO 16811:2014
3. Kleinert W., Oberdörfer Y.: Calculated Bandwidth Dependent DGS and DAC Curves for Phased Array Sizing. Proceedings ECNDT, Prag (2014). http://www.ndt.net/events/ECNDT2014/app/content/Paper/165_Kleinert.pdf

Chapter 11
Frequency-Dependent Sound Attenuation

Abstract Sound attenuation is usually considered as a fixed value when DGS evaluation is applied. In reality, the sound attenuation is frequency-dependent. When a pulse travels in a test specimen with sound attenuation the spectrum and the center frequency of the pulse is changed with the traveled sound path. But even the bandwidth-dependent DGS diagram is calculated for a fixed center frequency. A method to derive and to consider the frequency-dependent sound attenuation is introduced to further improve the DGS accuracy.

The sound attenuation is considered as a fixed value when DGS or DAC curves are applied, refer to Chap. 2. But the sound attenuation is frequency-dependent in reality. The spectrum of the ultrasonic pulse is changed significantly in test pieces with high sound attenuation and when longer sound paths occur. The center frequency is shifted to lower values. When DGS or DAC curves are calculated the center frequency is considered as a fixed value even if the correction for the sound attenuation is applied [1]. Instead of just measuring the amplitudes of the V and W through transmissions, the resulting echoes can be transferred into the frequency domain by applying a Fast Fourier Transformation.

V and W through transmissions were measured on a test block with a thickness of $d_t = 50\,\mathrm{mm}$, using a 4 MHz *true*DGS® single element angle beam probe with an angle of incidence of $\beta = 45°$. The amplitude difference was measured with $\Delta G_{vw} = 5.4\mathrm{dB}$. Additionally, the echoes were transferred into the frequency domain, Fig. 11.1. As, unfortunately, the resulting spectra were given as bitmaps, the two bitmaps had to be digitized manually (Fig. 11.2).

The gain difference based on the different sound paths has to be determined. Since the center frequencies for the V and W through transmissions are very close together the bandwidth-dependent DGS diagram was calculated for the center frequency of the V through transmission. The sound paths s_v and s_w for the through transmissions are given by:

$$\begin{aligned} s_v &= \frac{d_t}{\cos\beta} = \frac{50\text{ mm}}{\cos\ 45°} = 70.7\,\mathrm{mm} \\ \Rightarrow\quad s_w &= 141.4\,\mathrm{mm} \end{aligned}$$

W. Kleinert, *Defect Sizing Using Non-destructive Ultrasonic Testing*,
DOI 10.1007/978-3-319-32836-2_11

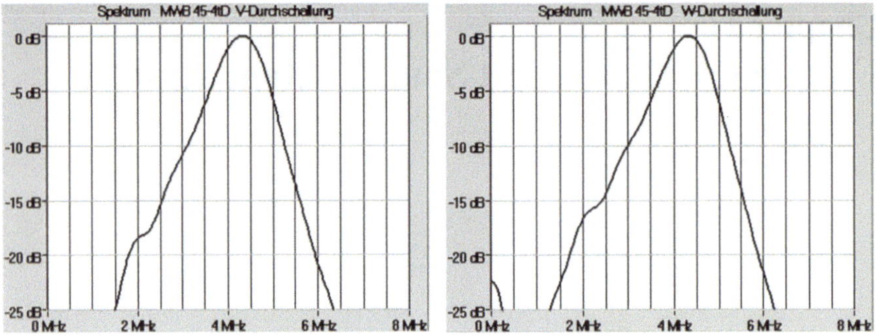

Fig. 11.1 Frequency spectrum for the V and W through transmission

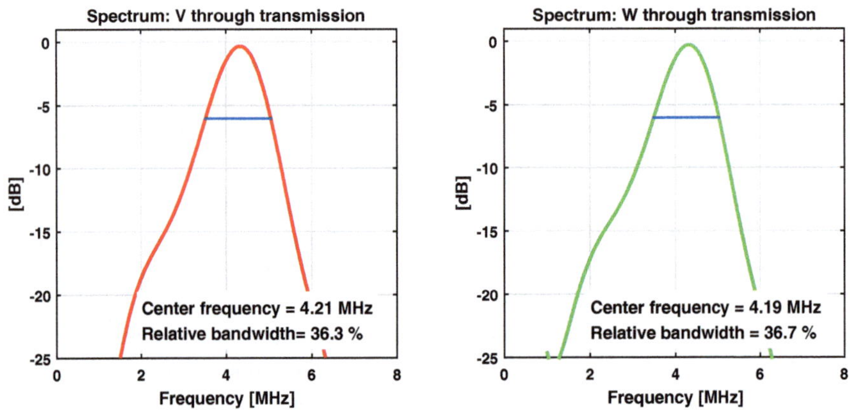

Fig. 11.2 Digitized frequency spectra for the V and W through transmissions

With these sound paths, the gain difference ΔG based on the different distances can be read from the back wall echo curve of the bandwidth-dependent DGS diagram, Fig. 11.3. For comparison, the sound attenuation κ is derived according to the state of the art without using the frequency domain first

$$\kappa = 1000 \, \frac{\Delta G_{vw} - \Delta G}{2 \, (s_w - s_v)} = 1000 \, \frac{1.6}{141.4} \, \frac{dB}{m} = 11.3 \, \frac{dB}{m} \tag{11.1}$$

To evaluate the sound attenuation frequency dependently the spectra are linearized and the spectrum of the W transmission is reduced by $\Delta G_{vw} - \Delta G$, Fig. 11.4. The ratio of the linear frequency amplitudes from the W and V trough transmissions results in linear values of the frequency-dependent sound attenuation. These values are converted to dB values. The result for the center frequency of the V through transmission is 10.7 dB/m (Fig. 11.5).

The original spectrum of the probe can be reconstructed when knowing the frequency-dependent sound attenuation. The compensation of the sound attenuation

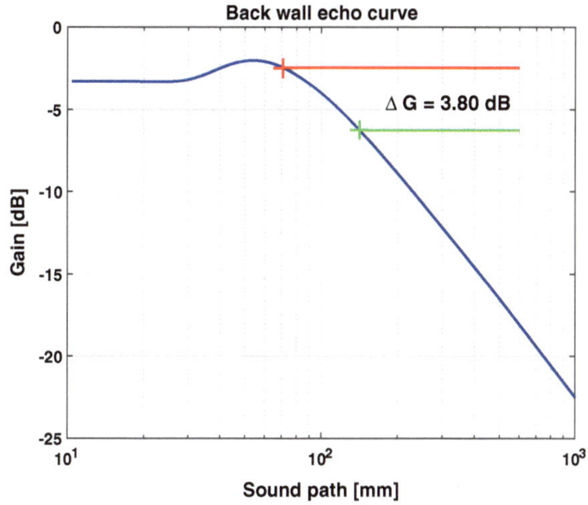

Fig. 11.3 Distance-based gain difference between the V and W through transmission

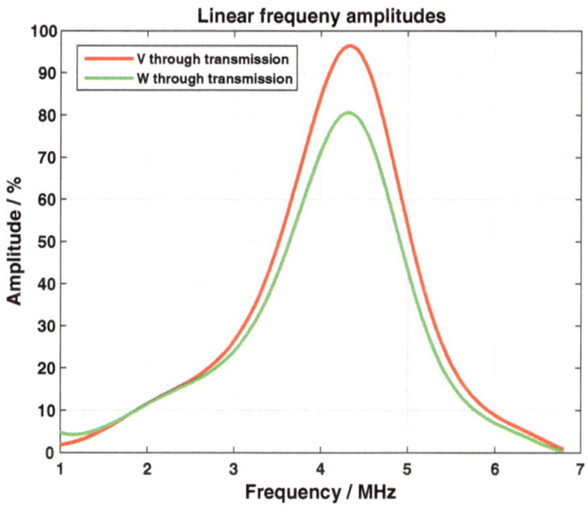

Fig. 11.4 Linear frequency amplitudes

in the spectrum for the V through transmission according to the sound path yields the original spectrum, Fig. 11.6. With the original spectrum and the frequency-dependent sound attenuation the spectrum for each sound path can be derived and considered for the DGS evaluation.

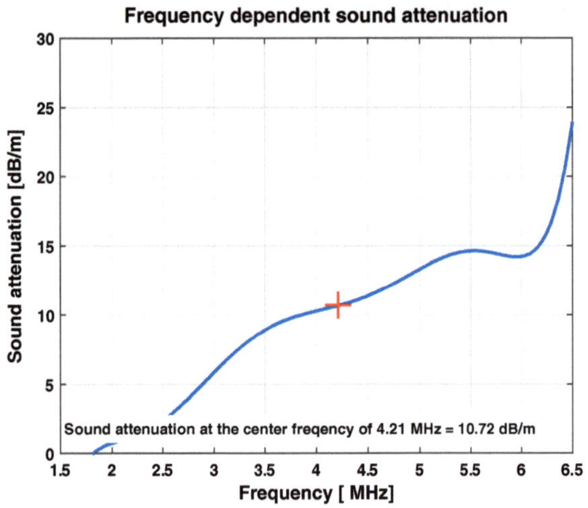

Fig. 11.5 Frequency-dependent sound attenuation

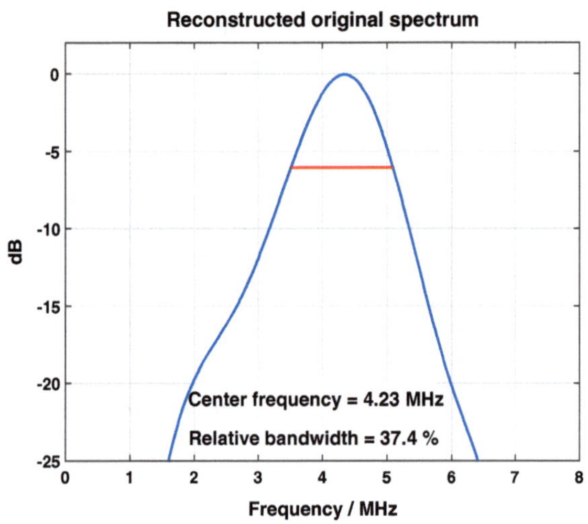

Fig. 11.6 Reconstructed original spectrum of the probe

Reference

1. Krautkramer, J., Krautkramer, H.: Ultrasonic Testing of Materials 4th Fully Revised Edition Translation of the 5th Revised German Edition Springer, Berlin GmbH. (1990)

Appendix

DGS Evaluation Taken with a 2 MHz Phased Array Probe

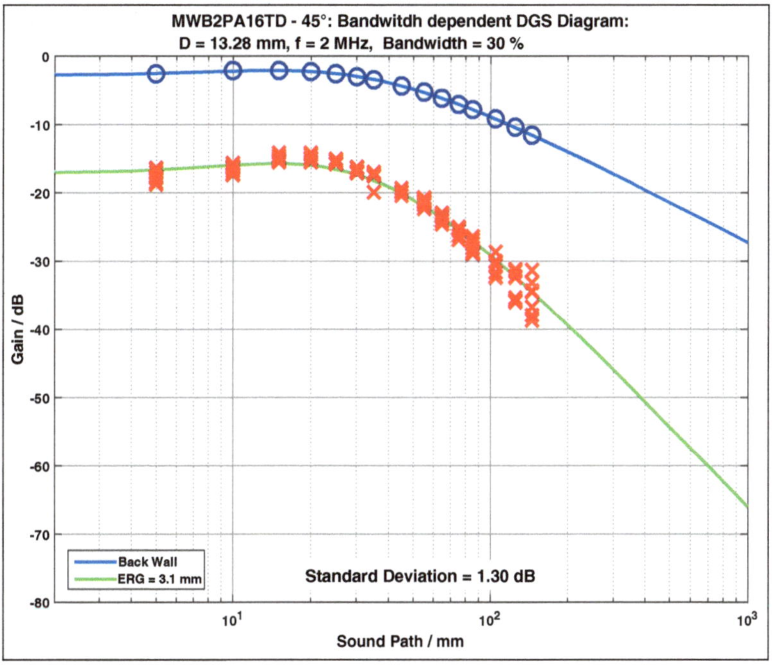

W. Kleinert, *Defect Sizing Using Non-destructive Ultrasonic Testing*,
DOI 10.1007/978-3-319-32836-2

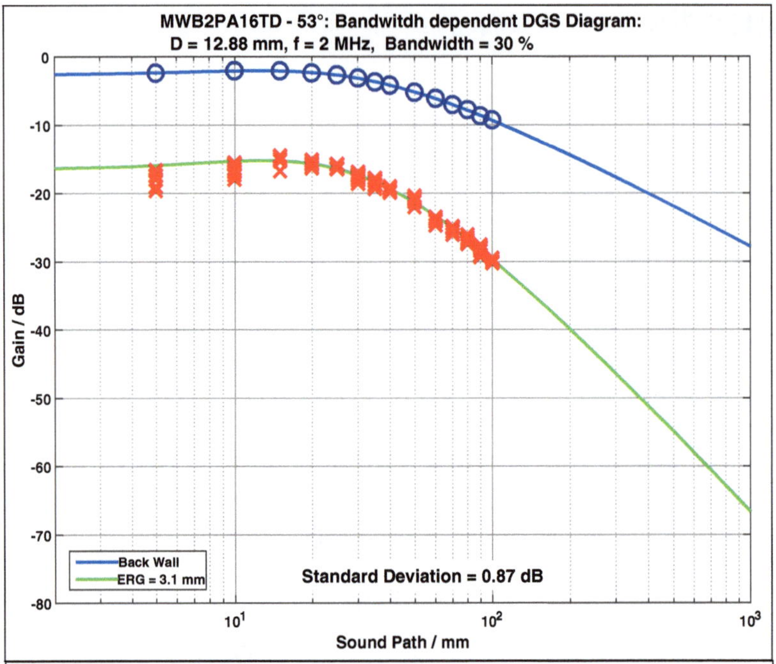

MWB2PA16TD - 53°: Bandwitdh dependent DGS Diagram:
D = 12.88 mm, f = 2 MHz, Bandwidth = 30 %
Gain / dB
Sound Path / mm
Back Wall
ERG = 3.1 mm
Standard Deviation = 0.87 dB

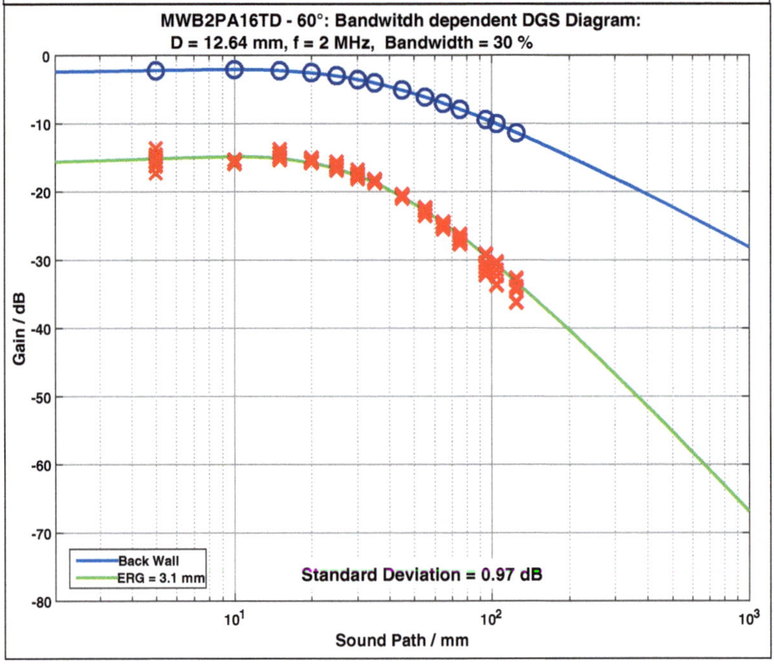

MWB2PA16TD - 60°: Bandwitdh dependent DGS Diagram:
D = 12.64 mm, f = 2 MHz, Bandwidth = 30 %
Gain / dB
Sound Path / mm
Back Wall
ERG = 3.1 mm
Standard Deviation = 0.97 dB

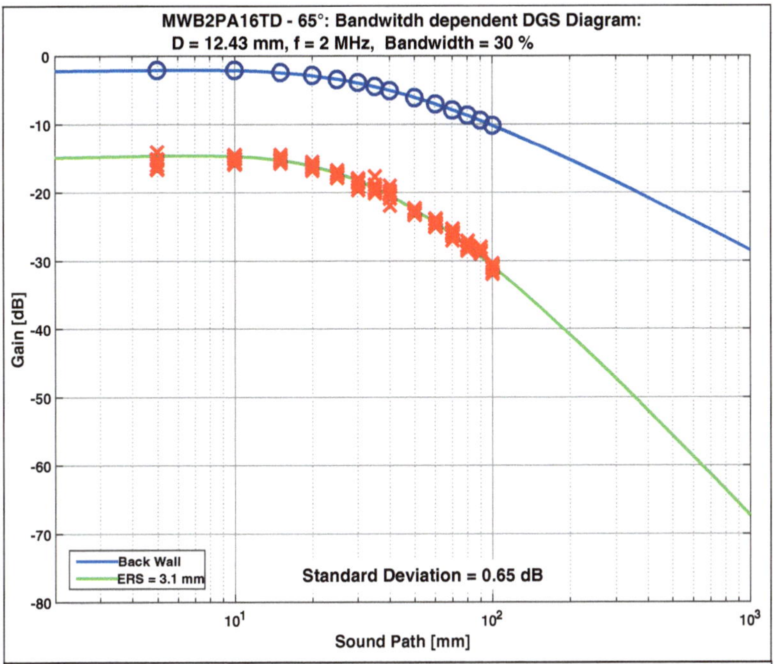
MWB2PA16TD - 65°: Bandwitdh dependent DGS Diagram:
D = 12.43 mm, f = 2 MHz, Bandwidth = 30 %
Gain [dB]
0
-10
-20
-30
-40
-50
-60
-70
-80
Back Wall
ERS = 3.1 mm
Standard Deviation = 0.65 dB
10^1
10^2
10^3
Sound Path [mm]

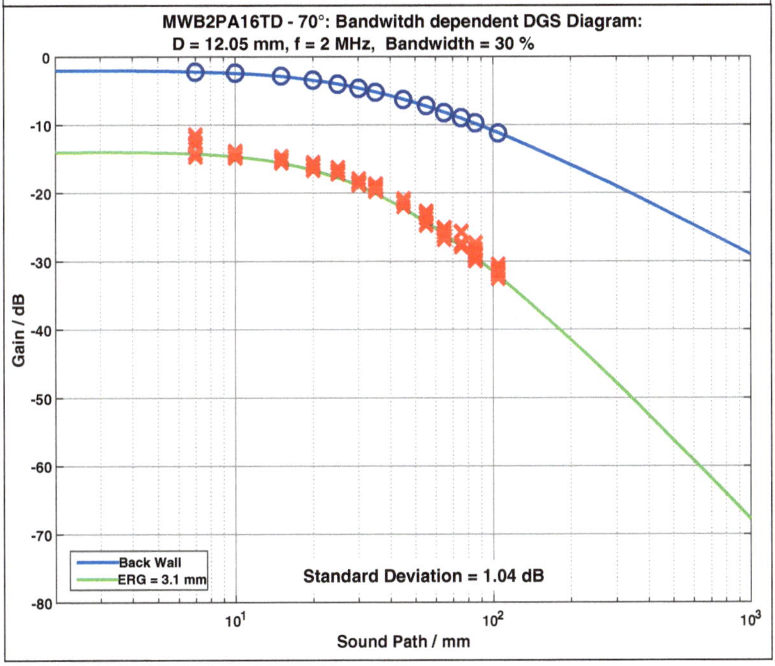
MWB2PA16TD - 70°: Bandwitdh dependent DGS Diagram:
D = 12.05 mm, f = 2 MHz, Bandwidth = 30 %
Gain / dB
0
-10
-20
-30
-40
-50
-60
-70
-80
Back Wall
ERG = 3.1 mm
Standard Deviation = 1.04 dB
10^1
10^2
10^3
Sound Path / mm

DGS Evaluation Taken with a 4 MHz Phased Array Probe

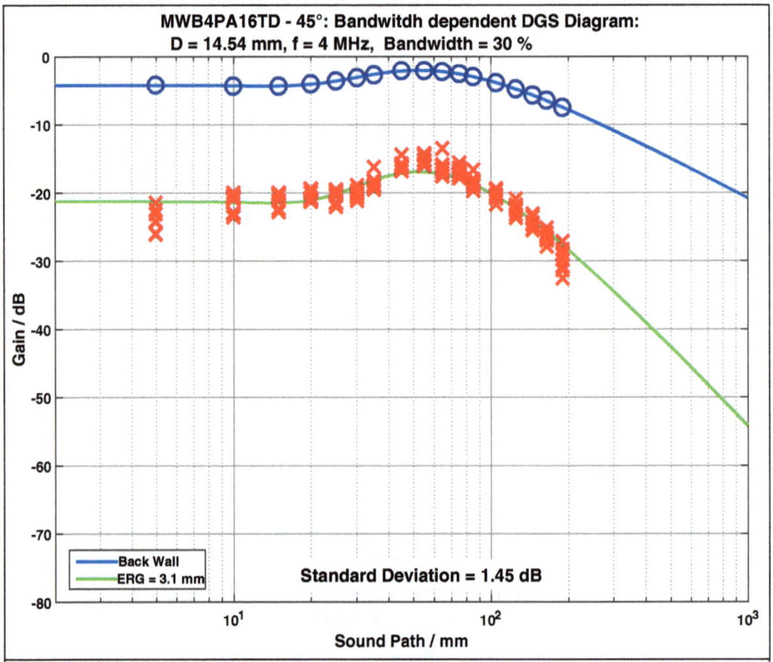

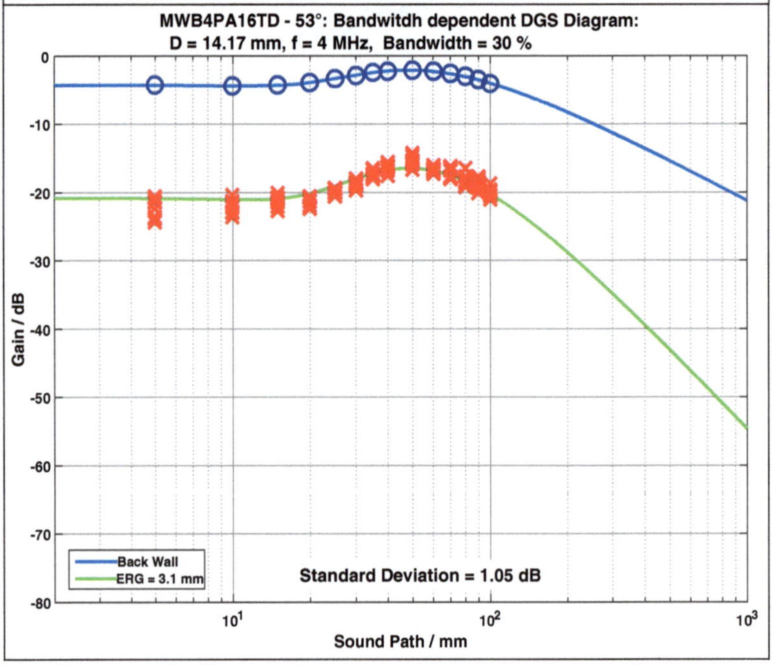

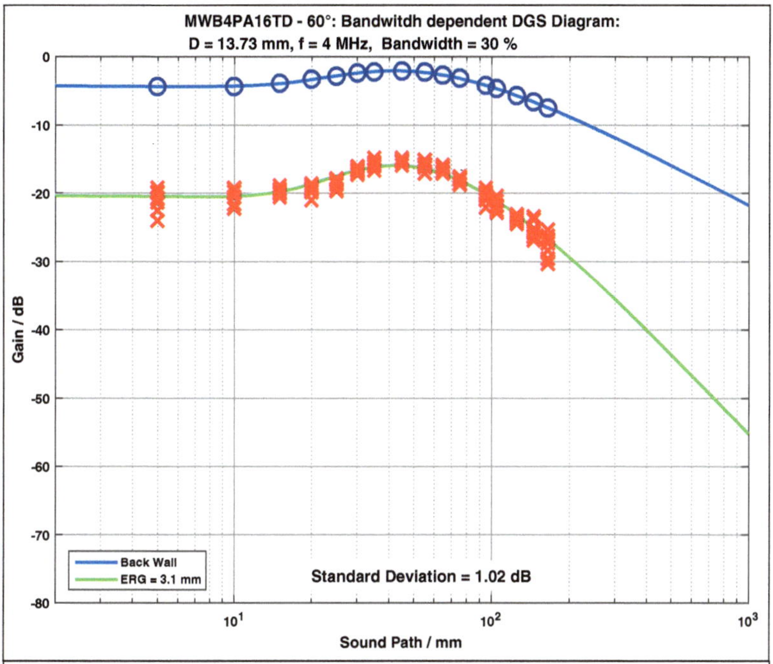
MWB4PA16TD - 60°: Bandwitdh dependent DGS Diagram:
D = 13.73 mm, f = 4 MHz, Bandwidth = 30 %
0
-10
-20
-30
-40
-50
-60
-70
-80
Gain / dB
Back Wall
ERG = 3.1 mm
Standard Deviation = 1.02 dB
10^1
10^2
10^3
Sound Path / mm

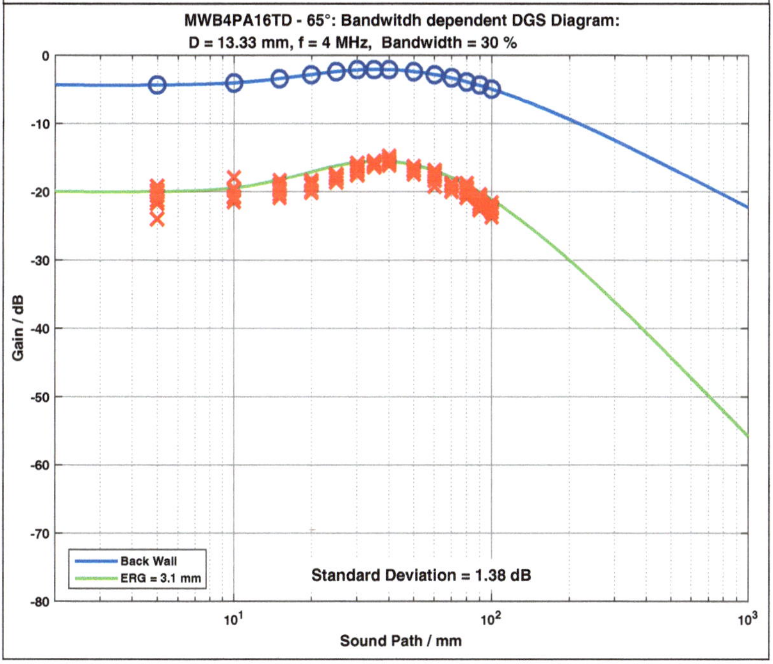
MWB4PA16TD - 65°: Bandwitdh dependent DGS Diagram:
D = 13.33 mm, f = 4 MHz, Bandwidth = 30 %
0
-10
-20
-30
-40
-50
-60
-70
-80
Gain / dB
Back Wall
ERG = 3.1 mm
Standard Deviation = 1.38 dB
10^1
10^2
10^3
Sound Path / mm

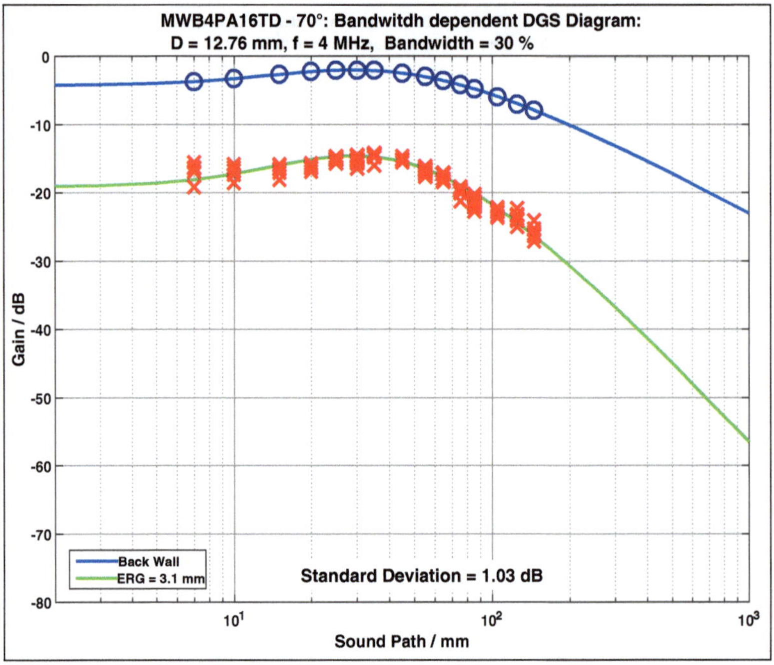
MWB4PA16TD - 70°: Bandwitdh dependent DGS Diagram:
D = 12.76 mm, f = 4 MHz, Bandwidth = 30 %
Gain / dB
0
-10
-20
-30
-40
-50
-60
-70
-80
Back Wall
ERG = 3.1 mm
Standard Deviation = 1.03 dB
10¹
10²
10³
Sound Path / mm

Further Readings

Literature

Schlengermann U.: *Zur Systematik der Entfernungsabhängigkeit des Druckes im Schallfeld von rechteckigen Ultraschallwandlern*, Deutsche Gesellschaft für Akustik (DAGA), VDI-Verlag, 1975, Seiten 441–444

Tietz, H. D.: *Ultraschall-Messtechnik*, VEB Verlag Technik Berlin, 1969

Wüstenberg H., Schulz E., Möhrle W., Kutzner J.: *Zur Auswahl der Membranformen bei Winkelprüfköpfen für die Ultraschallprüfung*, Materialprüfung 18, Nr. 7, Juli 1976, Seiten 223–230

Granted Patents and Patent Applications

EP 2 229 585 B1 *Method for the non-destructive testing of a test object by way of ultrasound and apparatus therefor*, GE Sensing & Inspection Technologies GmbH, 50354 Hürth (DE) W.-D. Kleinert, Y. Oberdörfer

EP 2 229 586 B1 *Method for the non-destructive testing of a test object using ultrasound, and apparatus therefor*, GE Sensing & Inspection Technologies GmbH, 50354 Hürth (DE) W.-D. Kleinert, Y. Oberdörfer

WO 2010/130819 A2 *Test probe as well family of test probes for the non-destructive testing of a workpiece by means of ultrasonic sound and testing device*, GE Sensing & Inspection Technologies GmbH, 50354 Hürth (DE) W.-D. Kleinert, G. Splitt

DE 10 2014 101 227 A1 *Vorrichtung und Verfahren zur zerstörungsfreien Prüfung eines Prüflings mittes Ultraschall nach der AVG-Methode*, GE Sensing & Inspection Technologies GmbH, 50354 Hürth (DE) W.-D. Kleinert

DE 10 2014 104 914 A1 *Vorrichtung und Verfahren zur zerstörungsfreien Prüfung mittels Ultraschall nach der Vergleichskörpermethode*, GE Sensing & Inspection Technologies GmbH, 50354 Hürth (DE) W.-D. Kleinert

W. Kleinert, *Defect Sizing Using Non-destructive Ultrasonic Testing*,
DOI 10.1007/978-3-319-32836-2

DE 10 2014 104 909 A1 *Verfahren und Vorrichtung zur zerstörungsfreien Prüfung eines Prüflings mittels Ultraschall unter Berücksichtigung der frequenzabhängigen Schallschwächung*, GE Sensing & Inspection Technologies GmbH, 50354 Hürth (DE) W.-D. Kleinert

US 5,511,425 *Flaw Detector incorporating DGS*, Krautkramer-Branson Inc. W.-D. Kleinert et. al.

Index

W. Kleinert, *Defect Sizing Using Non-destructive Ultrasonic Testing*,
DOI 10.1007/978-3-319-32836-2

MIX
Papier aus verantwortungsvollen Quellen
Paper from responsible sources
FSC® C105338

If you have any concerns about our products,
you can contact us on
ProductSafety@springernature.com

In case Publisher is established outside the EU,
the EU authorized representative is:
Springer Nature Customer Service Center GmbH
Europaplatz 3, 69115 Heidelberg, Germany

Printed by Libri Plureos GmbH
in Hamburg, Germany